Artificial Intelligence Trends for Data Analytics Using Machine Learning and Deep Learning Approaches

Artificial Intelligence (AI): Elementary to Advanced Practices

Series Editors: Vijender Kumar Solanki, Zhongyu (Joan) Lu, and Valentina E. Balas

In the emerging smart city technology and industries, the role of artificial intelligence is getting more prominent. This AI book series will aim to cover the latest AI work, which will help the naïve user to get support to solve existing problems and for the experienced AI practitioners, it will assist to shedding light for new avenues in the AI domains. The series will cover the recent work carried out in AI and its associated domains, it will cover logics, pattern recognition, NLP, expert systems, machine learning, block-chain, and big data. The work domain of AI is quite deep, so it will be covering the latest trends which are evolving with the concepts of AI and it will be helping those new to the field, practitioners, students, as well as researchers to gain some new insights.

Cyber Defense Mechanisms
Security, Privacy, and Challenges
Edited by Gautam Kumar, Dinesh Kumar Saini, and Nguyen Ha Huy Cuong

Artificial Intelligence Trends for Data Analytics Using Machine Learning and Deep Learning Approaches
Edited by K. Gayathri Devi, Mamata Rath, and Nguyen Thi Dieu Linh

For more information on this series, please visit: https://www.crcpress.com/Artificial-Intelligence-AI-Elementary-to-Advanced-Practices/book-series/CRCAIEAP

Artificial Intelligence Trends for Data Analytics Using Machine Learning and Deep Learning Approaches

Edited by

K. Gayathri Devi, Mamata Rath, and Nguyen Thi Dieu Linh

CRC Press
Taylor & Francis Group
Boca Raton London New York

CRC Press is an imprint of the
Taylor & Francis Group, an **informa** business

First edition published 2021
by CRC Press
6000 Broken Sound Parkway NW, Suite 300, Boca Raton, FL 33487-2742
and by CRC Press
2 Park Square, Milton Park, Abingdon, Oxon, OX14 4RN

© 2021 Taylor & Francis Group, LLC
CRC Press is an imprint of Taylor & Francis Group, LLC

Library of Congress Cataloging-in-Publication Data

Names: Gayathri Devi, K., editor. | Rath, Mamata, editor. | Dieu, Nguyen Thi, editor.
Title: Artificial intelligence trends for data analytics using machine learning and deep learning approaches / edited by K. Gayathri Devi, Mamata Rath and Nguyen Thi Dieu.
Description: Boca Raton, FL : CRC Press, 2021. | Series: Artificial intelligence (AI). Elementary to advanced practices | Includes bibliographical references and index. | Summary: "This book focuses on the implementation of various elementary and advanced approaches in AI that can be used in various domains to solve real-time decision-making problems"-- Provided by publisher.
Identifiers: LCCN 2020016370 (print) | LCCN 2020016371 (ebook) | ISBN 9780367417277 (hardback) | ISBN 9780367854737 (ebook)
Subjects: LCSH: Artificial intelligence--Industrial applications. | Machine learning--Industrial applications. | Diagnosis--Data processing.
Classification: LCC Q334.5 .A78 2021 (print) | LCC Q334.5 (ebook) | DDC 006.3--dc23
LC record available at https://lccn.loc.gov/2020016370
LC ebook record available at https://lccn.loc.gov/2020016371

ISBN: 978-0-367-41727-7 (hbk)
ISBN: 978-0-367-85473-7 (ebk)

Typeset in Times LT Std
by Cenveo® Publisher Services

Contents

Preface

Artificial intelligence (AI), when incorporated with machine learning and deep learning algorithms, has a variety of applications in today's society. Machine learning is an incredible breakthrough in the field of AI in which programs are developed to execute specific tasks is being implemented in different areas like medical diagnosis, object recognition, natural language processing, robotics control, information security, and the analysis of remote sensing images. This book elaborates on the implementation of various elementary and advanced approaches in AI in various domains. The theme will be helpful for the researcher and many other users of these technologies in the processing of images and data analytics.

ORGANIZATION OF THE BOOK

This book consists of 14 book chapters implemented with AI and machine learning approaches and case studies to solve real-time decision-making problems.

Chapter 1, An Artificial Intelligence System Based Power Estimation Method for CMOS VLSI Circuits

This chapter focuses on providing an alternative solution to estimate the power dissipation of CMOS VLSI circuits, using statistical tools such as BPNN and ANFIS, and to overcome the drawback of time complexity through the simulation of complex circuits using traditional power estimator tool.

Chapter 2, Awareness Alert and Information Analysis in Social Media Networking Using Usage Analysis and Negotiable Approach

This chapter analyzes the information of people's views and opinions posted in social media. The proposed usage analysis and negotiable approach will analyze the usage of young-aged adults from the age category of 18 to 35 on any particular social media network. The user interested test rate (UITR) analysis algorithm is applied to rate the user view analysis, interested domain, posting of comments, and sharing contents. The final information rate report will conclude the positive usage of the social media and awareness alert for user.

Chapter 3, Object Detection and Tracking in Video Using Deep Learning Techniques: A Review

This chapter highlights the fundamentals of object tracking methods and challenges in visual tracking. It also presents the various feature extraction methods, object classification, and tracking methods that should be applied for the implementation of a system using deep learning techniques. Section 3.4 provides an introduction to machine and deep learning techniques.

Chapter 4, Fuzzy MCDM: Application in Disease Risk and Prediction

This chapter implements various fuzzy multi-criteria decision-making methods for disease prediction and determination of drug dosage. The medical history of the patient and their family is analyzed and a probabilistic value is generated for the disease that could occur. This chapter also discusses the future scope of neuro-fuzzy application in detail with end analysis and precise results.

Chapter 5, Deep Learning Approach to Predict and Grade Glaucoma from Fundus Images through Constitutional Neural Networks

The chapter presents a novel method to automatically detect and grade glaucoma based on the severity level from digital fundus images using 25-layer convolutional neural network approaches. The implemented system is able to enhance the accuracy and calculation of performance metrics and indicates that deep learning based assessment of digital fundus images can be used as a viable decision-support system clinically in large-scale glaucoma screening and detection programs.

Chapter 6, A Novel Method for Securing Cognitive Radio Communication Network Using the Machine Learning Schemes and a Rule Based Approaches

This chapter proposes a novel machine learning method called "improved-apriori" algorithm to solve the sensing data falsification attack to enhance the security of 5G-based cognitive radio communication networks. The results obtained with the proposed work were compared with five existing approaches.

Chapter 7, Detection of Retinopathy of Prematurity Using Convolution Neural Network

This chapter focuses on the detection of retinopathy of prematurity (ROP) using a convolution neural network algorithm. In the proposed method the computational time for identification, batch loss during the process, and the accuracy were estimated for different sizes of convolutional layer.

Chapter 8, Impact of Technology on Human Resource Information System and Achieving Business Intelligence in Organizations

This chapter describes the historical evolution of the concept of human resource information system and the issues related to the handling of data analytics. It is a theoretical perspective to the concerned area related to the human resource information system. This chapter will help to identify the areas of research needed in implementing human resource information system in an organization and the achievement of business intelligence.

Chapter 9, Proficient Prediction of Acute Lymphoblastic Leukemia Using Machine Learning Algorithm

An automated leukocyte detecting system for the detection of acute lymphoblastic leukemia implemented with support vector machine technique is elaborated in this chapter. The improved accuracy of the proposed system was achieved by preprocessing techniques, clustering, feature extraction, and data classification with support vector machine technique. The combination of these techniques have been well experimented, executed, and justified for clinical diagnosis with the achieved results.

Chapter 10, Role of Machine Learning in Social Area Networks

This chapter uses machine learning concepts to resolve the data collected and maintained from different social sources. Rule-based algorithms is a learning classifier system that acts as a supervised machine learning system model for social media analysis. The flexible set of features resulted in a system that automates, collects, maintains, and manipulates the data without human intervention.

Chapter 11, Breast Cancer and Machine Learning: Interactive Breast Cancer Prediction Using Naive Bayes Algorithm

The machine learning classifier naive Bayes classifier is implemented in this chapter to determine the stage of a breast cancer patient and to grade the cell size.

The tumor, node, metastasis system is used to explore the stages of cancer. Accurate performance of the naive Bayes algorithm is evaluated by calculating the accuracy, specificity, sensitivity, and F1 score.

Chapter 12, Deep Networks and Deep Learning Algorithms

This chapter provides an overview of the concept and the ever-expanding advantages and popularity of deep learning. It highlights that the computational models developed using deep learning consist of multiple processing layers that are capable of representing the data with multiple levels of abstraction.

Chapter 13, Machine Learning for Big Data Analytics, Interactive and Reinforcement

The objective of this chapter is to provide a concise view on machine learning for big data analytics and to enlighten the reader about machine learning as interactive and reinforcement in it. This chapter provides a perspective on the domain, identifies research gaps and opportunities, and provides a strong foundation and encouragement in the field of machine learning for big data analytics.

Chapter 14, Fish Farm Monitoring System Using IoT and Machine Learning

This chapter has proposed a smart fishery checking framework with the assistance of the Internet of Things (IoT) to detect a favorable environment for the fish. This system has utilized quartz software and diverse sensors for water temperature, turbidity, pH, water level, CO, and NH3 gas. The enhanced outcomes obtained by the performance of the system are tested on a water environment.

CONCLUSION

The book reviews deep learning algorithms and their application for the detection of the object and tracking to run tasks in an automated manner. There are chapters that focus on the power estimation method for CMOS VLSI circuits using AI systems. it also provides insight on the big data analytics, social media networks, and effective cognitive radio communication using machine learning algorithms.

AI, in association with machine learning techniques, can develop incredible systems for the service of mankind, especially in the health care sector. Various case studies such as prediction of acute lymphoblastic leukemia, interactive breast cancer prediction and identification of ROP implemented using machine learning and deep learning algorithms are discussed in this book for the development of modern health care systems for accurate medical diagnosis and treatment procedures for immediate recovery. The system for the monitoring of fish farm in real-time scenario using IoT and machine learning approaches are also very useful.

TARGET AUDIENCE

This book aims to bring together leading academic scientists, researchers, and research scholars to exchange and share their experiences and research results on all aspects of machine learning and AI approaches. It also provides a premier interdisciplinary platform for researchers, practitioners, and educators to present and discuss the most recent innovations, trends, and concerns as well as practical challenges encountered and solutions adopted in the fields of image processing and data analytics.

MATLAB® is a registered trademark of The MathWorks, Inc. For product information, please contact:
The MathWorks, Inc.
3 Apple Hill Drive
Natick, MA, 01760-2098 USA
Tel: 508-647-7000
Fax: 508-647-7001
E-mail: info@mathworks.com
Web: www.mathworks.com

Acknowledgment

We would like to express our deep gratitude to Dr. Vijender Kr. Solanki, Associate Professor, Department of Computer Science and Engineering, CMR Institute of Technology, Hyderabad, and Valentina E. Balas, Associate Professor, Department of Automatics and Applied Software, Aurel Vlaicu University of Arad, Romania, for the constant encouragement, support, and guidance for the completion of book. I am also grateful to the editorial and production team of CRC Press.

Editors

K. Gayathri Devi has 20 years of experience as a professor in the Department of Electronics and Communication Engineering, Dr. N.G.P Institute of Technology, Tamil Nadu, India. She has published papers in national and international journals and conferences. She has published patents. She is the reviewer of many SCI and Scopus indexed journals. She has received many proficiency awards and grants, and is a topper in the NPTEL online certification examination. She is the life member of ISTE, International Association of Engineers and Institute of Research Engineers and Doctors. Her research interests include Medical Image Processing, Internet of Things, Artificial Intelligence, and Embedded Systems.

Mamata Rath has about 13 years of experience as an assistant professor. She has many research publications, including SCI, Scopus, dblp, and Web of Science indexing in different international journal and conferences. Her research interests include Information Systems, Wireless Networks, Internet of Things, Computer Security, Real Time Systems, Ubiquitous Computing, Big Data and Artificial Intelligence. Currently she is an assistant professor of Information Technology at Birla Global University, Odisha, India.

Nguyen Thi Dieu Linh has more than 18 years of academic experience in electronics, IoT, Telecommunication, Big Data, and Artificial Intelligence. She has published more than 15 research articles in national and international journals, books, and conference proceedings. She is reviewer for *Information Technology Journal, Mobile Networks and Applications Journal*, and some international conferences. Now she is a head of Department of Electronics and Telecommunication Engineering, Faculty of Electronics Engineering, Hanoi University of Industry (HaUI), Vietnam.

Contributors

N.B. Ananthamoorthy
Hindusthan College of Engineering
and Technology
Coimbatore, Tamil Nadu, India

K.N. Apinaya Prethi
Department of Computer Science and
Engineering
Coimbatore Institute of Technology
Coimbatore, Tamil Nadu, India

Rajeswari Arumugam
Department of Electronics and
Communication Engineering
Coimbatore Institute of Technology
Coimbatore, Tamil Nadu, India

Premalatha Balasubramaniam
Department of ECE
Coimbatore Institute of Technology
Coimbatore, Tamil Nadu, India

Kishore Balasubramanian
Dr. Mahalingam College
of Engineering and Technology
Coimbatore, Tamil Nadu, India

Poongodi Chenniappan
Department of Electronics and
Communication Engineering
Bannari Amman Institute of Technology
Sathyamangalam, Tamil Nadu, India

Archana Choudhary
Birla School of Management
Birla Global University
Bhubaneswar, Odisha, India

Samuel Theodore Deepa
Department of Computer Science
Shri Shankarlal Sundarbai Shasun Jain
College for Women
Chennai, Tamil Nadu, India

Sharanika Dhal
Birla School of Management
Birla Global University
Bhubaneswar, Odisha, India

Deepa Dhanaskodi
Department of Electronics and
Communication Engineering
Bannari Amman Institute of
Technology
Sathyamangalam, Tamil Nadu, India

Mohammad Farhan Ferdous
Japan-Bangladesh Robotics &
Advanced Technology Research
Center (JBRATRC) Japan

Atapaka Thrilok Gayathri
Department of Computer Science
Shri Shankarlal Sundarbai Shasun Jain
College for Women
Chennai, Tamil Nadu, India

Meenu Gupta
Department of Computer Science
and Engineering
Chandigarh University
Chandigarh, Punjab, India

Mahmudul Hasan
Jahangirnagar University
Savar, Dhaka, Bangladesh

Rachna Jain
Bharati Vidyapeeth's College
of Engineering
Delhi, India

Ramesh Jayabalan
PSG College of Technology
Coimbatore, Tamil Nadu, India

Cynthia Joseph
Department of Electronics and
  Communication Engineering
Coimbatore Institute of Technology
Coimbatore, Tamil Nadu, India

Kavitha Kanagaraj
Department of Electronics
  and Communication
  Engineering
Kumaraguru College of Technology
Coimbatore, Tamil Nadu, India

Abhishek Kathuria
Bharati Vidyapeeth's College of
  Engineering
Delhi, India

Tannu Kumari
Department of Computer Science and
  Information Technology
C.V. Raman Global University
Bhubaneswar, Odisha, India

Antony Hyils Sharon Magdalene
PSG College of Technology
Coimbatore, Tamil Nadu, India

Anjana Mishra
Department of Computer Science and
  Information Technology
C.V. Raman Global University
Bhubaneswar, Odisha, India

Devanshi Mukhopadhyay
Bharati Vidyapeeth's College
  of Engineering
Delhi, India

Alamelu Muthukrishnan
Department of Information
  Technology
Kumaraguru College of
  Technology
Coimbatore, Tamil Nadu, India

S. Nithya
Department of Computer Science and
  Engineering
Coimbatore Institute of Technology
Coimbatore, Tamil Nadu, India

Manas Kumar Pal
Birla School of Management
Birla Global University
Bhubaneswar, Odisha, India

Ritwik Raj
Department of Computer Science and
  Information Technology
C.V. Raman Global University
Bhubaneswar, Odisha, India

Dhivya Praba Ramasamy
Department of Electronics and
  Communication Engineering
Kumaraguru College of Technology
Coimbatore, Tamil Nadu, India

Mamata Rath
Birla School of Management
Birla Global University
Bhubaneswar, Odisha, India

M. Sangeetha
Department of Information Technology
Coimbatore Institute of Technology
Coimbatore, Tamil Nadu, India

Lakshmanan Thulasimani
PSG College of Technology
Coimbatore, Tamil Nadu, India

Farjana Yeasmin Trisha
Japan-Bangladesh Robotics &
  Advanced Technology Research
  Center (JBRATRC) Japan

Govindaraj Vellingiri
Sri Venkateswara College of
  Engineering and Technology
Chitoor, Andhra Pradesh, India

1 An Artificial Intelligence System Based Power Estimation Method for CMOS VLSI Circuits

Govindaraj Vellingiri
Sri Venkateswara College of Engineering and Technology

Ramesh Jayabalan
PSG College of Technology

CONTENTS

1.1 INTRODUCTION

With the increased use of portable devices such as laptops, cellular phones, etc., power consumption has become a major issue that determines battery life span. Advances in very large scale integration (VLSI) technology have led to the fabrication of chips

1

that contain millions of transistors. In nanometer or deep submicron technology, power consumption has become an essential concern due to factors such as the increase in number of transistors on a chip and speed due to scaling of transistor size. Hence there is a need to minimize power dissipation [1]. Under these constraints, it is necessary to estimate the average power consumption during the design phase. This chapter explains how to employ artificial intelligence systems such as back-propagation neural network (BPNN) and adaptive neuro-fuzzy inference system (ANFIS), which have the capability to accurately estimate power for the CMOS VLSI circuits, without knowledge of circuit structure and interconnections.

1.1.1 Previous Work Using BPNN and ANFIS

Application of ANFIS for computing resonant frequency in microstrip antennas was discussed in the literature [2]. Al-Shammari et al [3] designed an ANFIS system to estimate wind farm wake effect. Bhanja and Ranganathan [4] proposed a Bayesian network based method of switching activity measurement in VLSI circuits that employs logic-induced directed acyclic graphing within a realistic time and accuracy. Simulation and non-simulation-based approaches of average power were discussed in literatures [5] and [6]. Application of BPNN to estimate power for combinational and sequential circuits is discussed in references [1, 7–9]. Karimi et al [10] extracted small-signal equivalent circuit model (S-parameter data) of bipolar transistors using ANFIS. Guney et al [11] discussed resonant frequency calculation for microstrip antennas using ANFIS. Güler et al [12] presented a new approach using ANFIS for coplanar waveguides gap discontinuities identification. Fault classification for PLL using BPNN was presented by Ramesh et al [13]. Application of ANFIS to predict air temperature to provide knowledge about climate and drought detection was discussed by Karthika et al [14], in which the authors compared and showed that the Gaussian membership function performs better than the Gbell membership in ANFIS. Estimation of power using BPNN that used the Levenberg-Marquardt function was proposed by Hou et al [15]. Mohammadi et al [16] estimated wind-power density by using a model based upon extreme learning machine (ELM), showing that BPNN can also be used for automating power estimation. Muragavel et al [17] discussed estimating average power through exhaustive simulation for larger input circuits. Nikolić et al [18] demonstrated application of ELM for sensor-less wind-speed predictions. Nikolić et al [19] explained wake power and wind speed deficit prediction using soft computing techniques. Nikolić et al [20] proposed an application of ANFIS to estimate wind turbine noise levels. Nikolić et al [21] discussed statistical analysis such as root mean square error (RMSE) and coefficient of determination (R) of wind speed using ANFIS. Applying ANFIS to wind-power modeling and wind turbines were discussed in reference [22]. ANFIS-based prediction of the modulation transfer function of an optical lens system was discussed in references [23] and [24]. Petković et al [25] proposed a continuously variable transmission-based wind generator using an ANFIS controller. Petković et al [26] discussed applying ANFIS to select variables and analyzing the wake effect of wind turbines, since an adjusted turbine can affect the wind turbulence.

Petković et al [27] developed a turbine model to simulate power and wind veloc-ity for different building geometrics. Petković et al [28] used ANFIS for wind speed and direction frequency dispersion. Petković et al [29] discussed predic-tion of annual wind speed probability density distribution using ANFIS. Petković [30] discussed how to use ANFIS to predict the probability density distribution of wind speed. Ramanathan et al [31] proposed a power estimation technique using BPNN and a radial basis function neural network of ISCAS'89 benchmark circuits in which only two training functions, Traingdm and Traingdx, are used. Lorenzo at al [32] discussed leakage power minimization at the circuit level using SPICE tool. Shamshirband et al [33] and [34] investigated the application of ANFIS to predict the wind wake added turbulence. Shamshirband et al [35] discussed determining wind turbine noise levels due to wind speed and sound using ANFIS. Shamshirband et al [36] discussed application of ANFIS to predict the wake power deficit. Shamshirband et al [37] analyzed three wind-speed mod-els and applied ANFIS to determine the best mode. Govindaraj et al discussed power estimation using ANFIS in reference [38].

From the works discussed above, the drawbacks of the existing power estimation methods using BPNN are:

1. If the circuit size is large, defining the vector set for simulation will be too large or too difficult in simulation-based methods.
2. The simulation and probabilistic methods increase resource complexity and take more time since they require the complete structure of the VLSI circuit and interconnections.

ANFIS can be applied to predict wind speed, suitable wind turbines, and resonant frequency in antennas, and it is well suited for power estimation applications for CMOS VLSI circuits since it has low RMSE and high R. Power estimation using BPNN is done by changing various parameter such as the learning rate, momen-tum constant, hidden layer, and epoch, with 11 different algorithms. Power estima-tion using ANFIS is done by varying error goal, step size, membership function, and step-size increase rate and decrease rate. The error percentage between actual power and power obtained from the network is calculated for both BPNN and ANFIS.

1.2 TRAINING AND TESTING DATA

The ISCAS'89 benchmark circuit database has been considered for training and test-ing ANFIS and neural networks. It is obtained from reference [1], which is shown in Tables 1.1 and 1.2. The ISCAS'89 benchmark circuits consist of 25 sets of data, out of which 20 sets are used to train the proposed BPNN and ANFIS and five sets of data are used for testing the network. The database of the ISCAS'89 sequential circuits includes information regarding attributes such as D flip-flops, NAND gates, inverters, NOR gates, AND gates, OR gates, number of inputs, number of outputs, and total number of gates in the circuit.

TABLE 1.1

ISCAS'89 Benchmark Circuits Testing Data Set of ANFIS/Neural Network [1]

Benchmark Circuit Classification	Number of Inputs	Number of Outputs	Number of D Flip-Flops (D FF)	Number of Inverters	Total Number of Gates	Number of AND Gates	Number of NAND Gates	Number of OR Gates	Number of NOR Gates
S641	35	24	19	272	107	90	4	13	0
S344	9	11	15	59	101	44	18	9	30
S386	7	7	6	41	118	83	0	35	0
S382	3	6	21	59	99	11	30	24	34
S400	3	6	21	58	106	11	36	25	34

TABLE 1.2

ISCAS'89 Benchmark Circuits Training Data Set of ANFIS/Neural Network [1]

Benchmark Circuit Classification	Number of Inputs	Number of Outputs	Number of D Flip-Flops (D FF)	Number of Inverters	Total Number of Gates	Number of AND Gates	Number of NAND Gates	Number of OR Gates	Number of NOR Gates	Monte Carlo Simulation Power in mw/MHz
S713	35	23	19	254	139	94	28	17	0	0.03743
S5378	35	49	179	1775	1004	0	0	239	765	0.23357
S35932	35	320	1728	3861	12204	4032	7020	1152	0	1.22048
S838	34	1	32	158	288	105	57	56	70	0.01292
S13207	31	121	669	5378	2573	1114	849	512	98	0.354
S38417	28	106	1636	13470	8709	4154	2050	226	2279	1.14518
S9234	19	22	228	3570	2027	955	528	431	113	0.28004
S420	18	1	16	78	160	49	29	28	34	0.00903
S820	18	19	5	33	256	76	54	60	66	0.02831
S1423	17	5	74	167	490	197	64	137	92	0.07181
S953	16	23	29	84	311	49	114	36	112	0.02458
S1238	14	14	18	80	428	134	125	112	57	0.06347
S15850	14	87	597	6324	3448	1619	968	710	151	0.51991
S208	10	1	8	38	66	21	15	14	16	0.00698
S38584	12	278	1452	7805	11448	5516	2126	2621	1185	1.87987
S349	9	11	15	57	104	44	19	10	31	0.01856
S1488	8	19	6	103	550	350	0	200	0	0.0564
S1494	8	19	6	89	558	354	0	204	0	0.06018
S298	3	6	14	44	75	31	9	16	19	0.00912
S444	3	6	21	62	119	13	58	14	34	0.01172

1.3   POWER ESTIMATION USING A NEURAL NETWORK

The method consists of the following steps:

A. Construction of neural network
B. Training phase
C. Testing phase

1.3.1   CONSTRUCTION OF A NEURAL NETWORK

A back-propagation (BP) feed-forward network with four layers is constructed. The first layer transfer function is set as "linear" and the rest of the layers are set with "tansig" function. The parameters of BPNN such as hidden layer, learning rate, epoch, and momentum constant are varied for 11 different training algorithms such as Traingd, Traingda, Trainrp, Traingdx, Traingdm, Traincgf, Traincgp, Traincgb, Trainscg, Trainbfg, and Trainoss. The number of inputs chosen for sequential circuits is nine, since it has nine attributes which is shown in Figure 1.1. The number of attributes is varied from nine to six by eliminating OR gates, AND gates, and inverters.

1.3.2   BPNN TRAINING PHASE

Step 1: To train the BPNN, two-thirds of the input vectors are extracted from ISCAS'89 benchmark circuits' database [39].
Step 2: Normalization is done between −1 and +1 for input vectors and their target vectors since the tan sigmoidal activation function is used for all the hidden layers.
Step 3: After normalization, input vectors and their corresponding normalized target vectors are used to train the neural network.

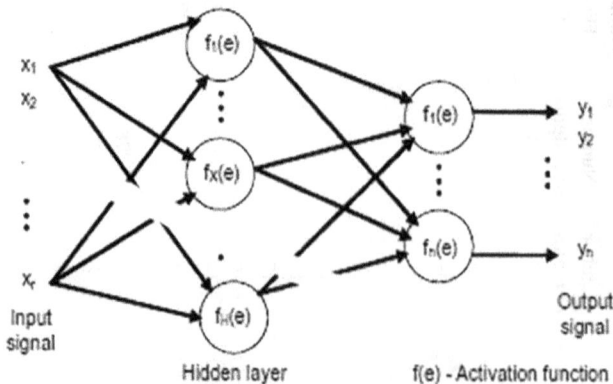

FIGURE 1.1   General architecture of BPNN.

TABLE 1.3
Range of BPNN Parameters

Parameter	Range
Hidden layer neurons	10–17
Momentum constant	0.1–0.9
Learning rate	0.3–0.8
Epochs	150–2700

1.3.3 BPNN TESTING PHASE

Step 1: To test the BPNN, one-third of the input vectors are used [39].

Step 2: Normalization is done for input vectors before testing.

Step 3: Output vectors for the corresponding normalized test input vectors are normalized and applied to BPNN.

Step 4: After testing, the original values of the output vectors are obtained by using a reverse normalization process.

Step 5: Validation for output vectors are done by performing regression analysis.

The BPNN-based training and testing process flow is illustrated in Figure 1.2.

1.3.4 NETWORK PARAMETERS

The BPNN is trained with 11 different training algorithms. The performance of these training algorithms is compared by varying the network parameters like learning rate, momentum constant, hidden layer, and epoch. Extensive MATLAB simulation is done to select the stopping criteria. The parameters are varied in the range as given in Table 1.3.

1.4 PROPOSED POWER ESTIMATION USING ANFIS TECHNIQUE

1.4.1 OVERVIEW OF THE PROPOSED WORK

Application of ANFIS is proposed to estimate power in CMOS VLSI circuits by varying different circuit attributes such as D flip-flops, NAND gates, inverters, NOR gates, AND gates, OR gates, number of inputs, number of outputs, and total number of gates in the circuit of ISCAS'89 benchmark. ANFIS is a combination of both fuzzy inference system and artificial neural network, which is its main advantage. Error, usually the sum of the square difference between actual output and desired output, is reduced at each iteration. Parameters of ANFIS, such as step size, error goal, membership function, step-size increase rate, and decrease rate, are varied to estimate power.

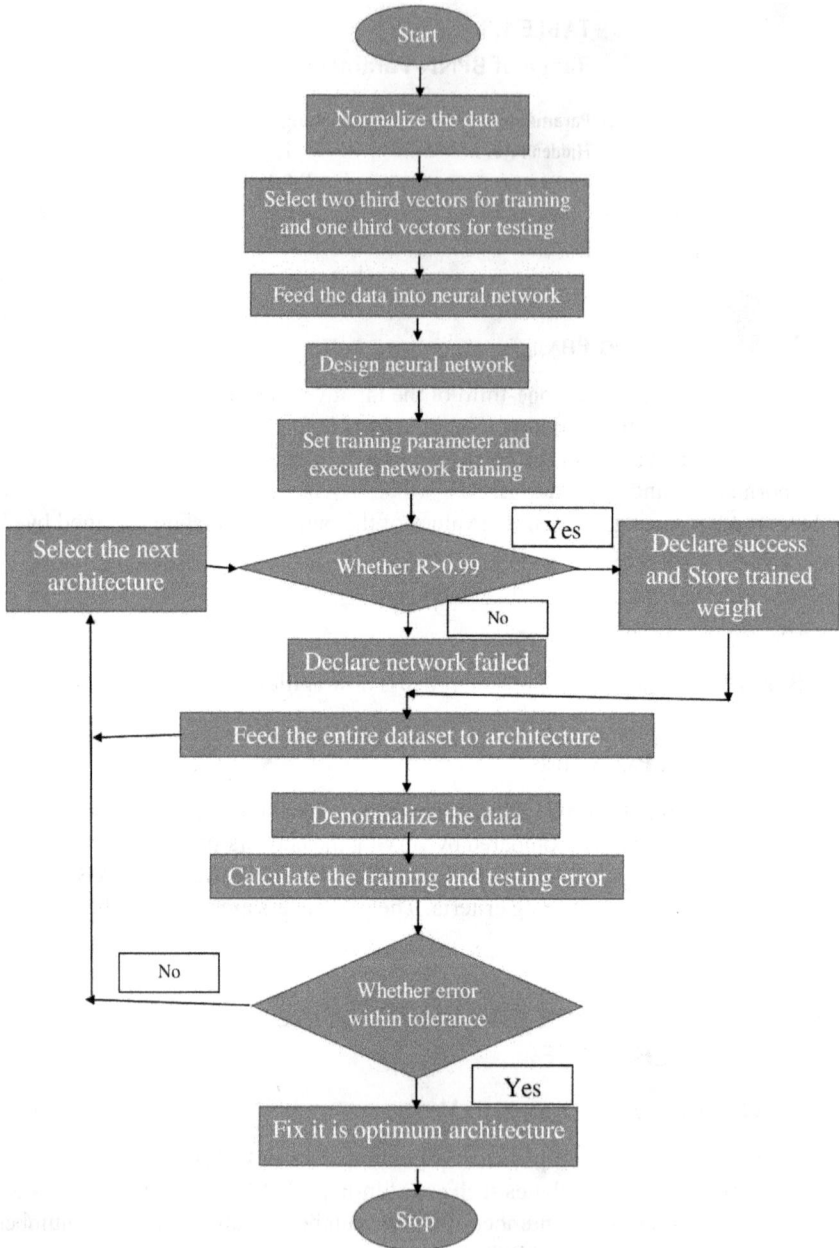

FIGURE 1.2   Proposed BPNN workflow for power estimation.

TABLE 1.4
ANFIS Range of Parameters

Parameters	Value/Range
Error goal	6
Initial step size	0.01
Step-size decrease rate	0.9
Step-size increase rate	1.5
Membership function	GAUSSIAN/SIGMF
Optimization method	HYBRID/BACK PROPAGATION

1.4.2 Training and Checking Used in ANFIS

Subtractive clustering technique and grid partitioning technique are used to frame the fuzzy rules. Initial FIS models are generated using an input configuration that was found previously. Hybrid learning algorithms are used to train the network and to tune the membership functions of the models obtained previously to provide as many accurate estimates as possible. The ANFIS training/checking flow chart is shown in Figure 1.3. The data set used for training/checking is shown in Tables 1.1 and 1.2.

1.4.3 Designing the ANFIS

An ANFIS consists of five layers that are used to tune the parameters in an FIS with hybrid learning mode and to implement different nodal functions. MATLAB tool R2010A is used to create an ANFIS model. Power estimation using ANFIS is done by varying step size, error goal, membership function, step-size increase rate and decrease rate. The ANFIS parameter variations range is shown in Table 1.4. Exhaustive MATLAB simulation is done to select the parameters for which the ANFIS performs well. ANFIS architecture for power estimation of ISCAS'89 circuit with 9 inputs and 10 fuzzy rules is shown in Figure 1.4.

1.5 RESULTS AND DISCUSSIONS

1.5.1 BPNN-Based Method

A power estimation technique based on BPNN is proposed for ISCAS'89 sequential circuits. The neural network is trained with 11 different training algorithms. The BPNN is trained six to nine attributes for sequential logic circuits. Performance of Trainscg deviates from ideal power consumption only by 0.01% for sequential circuits (ISCAS'89). A comparative study on power estimation is

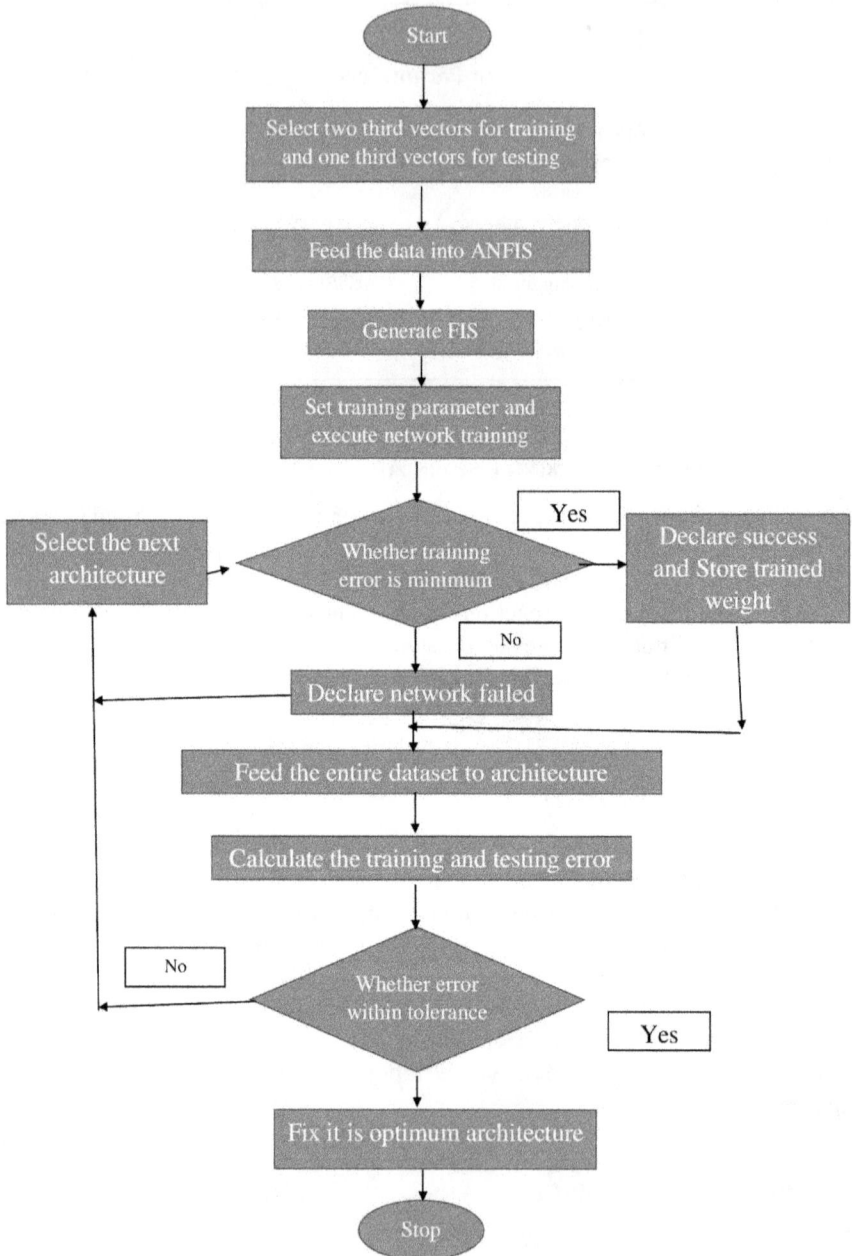

FIGURE 1.3   Proposed ANFIS workflow for power estimation.

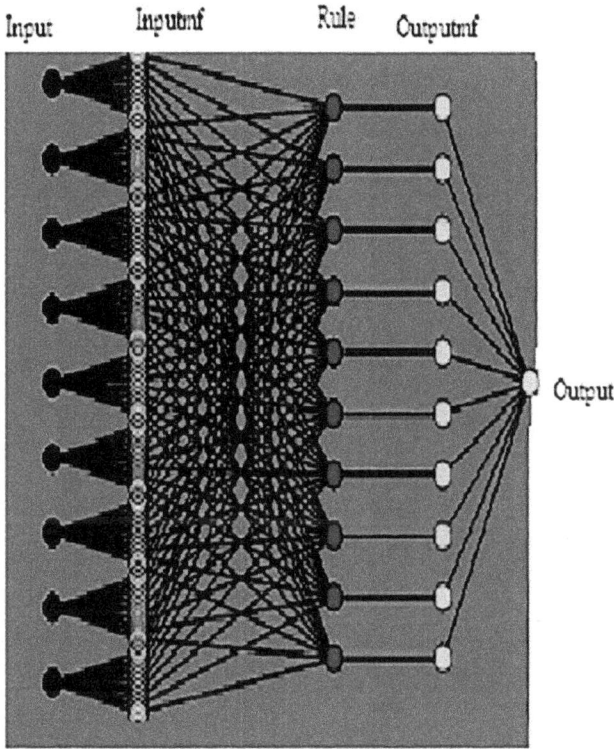

FIGURE 1.4   ANFIS architecture for power estimation of ISCAS'89 circuits.

made and inference is tabulated for various training algorithms. Table 1.5 shows the regression results for ISCAS'89 benchmark circuits. Power estimation using ANFIS is done for ISCAS'89 sequential circuits and it can be extended for other circuits also.

1.5.2   CALCULATING PREDICTION ERROR

Calculation of prediction error during testing and training is done by

$$\text{Error } \% = \frac{A - B \times 100}{A} \tag{1.1}$$

Where A is the actual value estimated using Monte Carlo simulation and B is the predicted value by testing the network.

Table 1.5 reports the regression analysis in which Trainscg for ISCAS'89 sequential circuits of layer size of 9:15:15:1 with 9 inputs and 253 epoch under conjugate gradient category is well suited for the power estimation.

Among the 11 different BPNN training algorithms, Trainscg updates the weight and bias values in a network training function according to the scaled conjugate

TABLE 1.5

Regression Results of ISCAS'89 Benchmark Circuits

Training Function	Number of Input Attributes	Network Layer Size	Number of Epochs	Slope (M)	Y-intercept (B)	Regression Value (R)
Traingdm	9	8:13:14:1	400	0.015	0.0016	0.992
Traingd	9	9:15:15:1	750	1.8	−0.027	0.994
Traingda	9	9:15:14:1	500	0.58	0.011	0.9956
Traingdx	9	8:14:15:1	225	0.34	0.018	0.9979
Traincgp	9	7:15:15:1	300	1.8	−0.025	0.9988
Traincgf	9	7:15:15:1	256	1.1	−0.0036	0.999
Trainoss	9	8:14:15:1	600	0.99	0.0017	0.9991
Trainbfg	9	8:14:15:1	600	0.99	0.0017	0.9991
Traincgb	9	9:16:15:1	100	1.7	−0.023	0.9993
Trainrp	9	8:16:15:1	250	1	-1.6×10^{-5}	0.9998
Trainscg	9	9:15:15:1	253	1	−0.0034	0.9999

gradient method. Trainscg is well suited for a network circuit with a large number of inputs and has a fast function approximation. Even when the error is reduced Trainscg performance does not degrade considerably as in the case of other algorithms. Therefore, Trainscg gives better result in terms of regression, number of epoch, Y-intercept (B), and slope(M) when compared to the existing methods as shown in Table 1.6.

1.5.3 ANFIS-Based Method

Data used for input and checking are loaded in the workspace and imported to the ANFIS tool in MATLAB. A new FIS is generated by performing exhaustive

TABLE 1.6

Regression Results Comparison of Existing Methods for ISCAS '89 Circuits Using BPNN

Training Function	Activation Function	Number of Input Attributes	Network Layer Size	Number of Epochs	Slope (M)	Y-Intercept (B)	Regression Value (R)
Trainscg	Tan sig	9	9:15:15:1	253	1	−0.0034	0.9999
Trainbfg	Tan sig	9	9:15:15:1	500	1.4	−0.016	0.9993
Trainlm [1]	Tan sig	9	9:15:15:1	878	0.464	−0.000882	0.753
Trainlm [1]	Tan sig	9	6:15:15:1	2139	0.915	−0.000963	0.994
Traingdx [30]	Tan sig	8	8:13:15:1	265	1.045	0.0051	0.9999
Traingdx [30]	log sig	8	7:14:15:1	225	1.044	2.77 E-04	0.9982

TABLE 1.7

Comparison of Power Estimation Using ANFIS with Gaussian Function

| | | Back Propagation | | | | Hybrid Optimization | | | |
| | | Constant Method | | Linear Method | | Constant Method | | Linear Method | |
Benchmark Circuit	Actual Power in mw	Obtained Power in mw	Error %	Obtained Power in mw	Error %	Obtained Power in mw	Error %	Obtained Power in mw	Error %
S344	0.01846	0.0182	1.41	0.0386	109.10	0.0185	0.22	0.0183	0.87
S382	0.01048	0.0112	6.87	0.0183	74.62	0.0107	2.10	0.0104	0.76
S386	0.0162	0.0112	30.86	0.0026	83.95	0.0162	0.00	0.0162	0.00
S400	0.01065	0.0195	83.10	0.0189	77.46	0.0107	0.47	0.0107	0.47
S641	0.03629	0.0146	59.77	0.0259	28.63	0.0385	6.09	0.0364	0.30

simulation and varying the ANFIS parameter in the range as shown in Table 1.4. An initial FIS model for power estimation applications using back-propagation and hybrid optimization with constant and linear method is created, in which minimum error is obtained using hybrid optimization with linear method configuration. Results for various ANFIS structures using Gaussian and sig member functions with testing error are reported in Tables 1.7 and 1.8.

Table 1.9 shows the power consumption comparison between ANFIS and BPNN, while Table 1.10 gives the details about BPNN and ANFIS error percentage. Among the 11 different BPNN algorithms, Trainscg function is best suited for ISCAS'89 sequential benchmark circuits. ANFIS, with hybrid learning and subtractive clustering, performs better than BPNN. ANFIS is a combination of fuzzy inference system and artificial neural network, which is its main advantage. Error, which is usually the sum of the square difference between actual output and desired output, is reduced at

TABLE 1.8

Comparison of Power Estimation Using ANFIS with Sig Member Function

| | | Back Propagation | | | | Hybrid Optimization | | | |
| | | Constant Method | | Linear Method | | Constant Method | | Linear Method | |
Benchmark Circuit	Actual Power in mw	Obtained Power in mw	Error %	Obtained Power in mw	Error %	Obtained Power in mw	Error %	Obtained Power in mw	Error %
S344	0.01846	0.0087	52.87	0.01048	43.23	0.0177	4.12	0.00656	64.7
S382	0.01048	0.0087	16.98	0.0184	75.57	0.0162	54.58	0.015	43.13
S386	0.0162	0.0162	0.00	0.0162	0.00	0.0107	33.95	0.0162	0.00
S400	0.01065	0.0105	1.41	0.0107	0.47	0.0177	66.20	0.0107	0.47
S641	0.03629	0.0087	76.03	0.029	20.09	0.0177	51.23	0.0668	84.07

TABLE 1.9

Comparison of Power Estimation Results of BPNN and ANFIS

Benchmark Circuit	Actual Power in mw	BPNN in mw	ANFIS in mw
S344	0.01846	0.0193	0.0183
S382	0.01046	0.0176	0.0104
S386	0.01628	0.0182	0.0162
S400	0.01065	0.01087	0.0107
S641	0.03629	0.0246	0.0364

each iteration. ANFIS has the capability to generate an FIS that gives a linear relationship between the input and output data. Hence, ANFIS' exclusive characteristics appear to be a better choice for estimation of power in CMOS VLSI circuits.

The ANFIS is a combination of fuzzy inference systems and artificial neural networks, so it has the advantage of more adaptation capability, rapid learning capacity, and the ability to capture the nonlinear structure of a process and this is why it is well suited to estimate the power of CMOS VLSI circuits which is shown in Figures 1.5 and 1.6. Figure 1.5 shows a graph that compares the actual power of the circuit and the power estimations obtained by testing the network using BPNN and ANFIS. From the graph we can infer that power estimated using ANFIS follows the actual power estimated by Monte Carlo simulation. Figure 1.6 gives the information about error percentage comparison in which ANFIS gives minimum error percentage when compared to BPNN.

1.5.4 PERFORMANCE EVALUATION

The RMSE and R are used to evaluate the performance of ANFIS and BPNN, which can be calculated by equations (1.2) and (1.3)

$$RMSE = \sqrt{\frac{\sum_{i=1}^{N}\left(Y_I^O - Y_I^C\right)^2}{N}} \tag{1.2}$$

TABLE 1.10

Comparison of Error Calculation of BPNN and ANFIS

Benchmark Circuit	BPNN (%)	ANFIS (%)
S344	4.55	0.87
S382	67.93	0.76
S386	12.34	0.00
S400	2.06	0.47
S641	32.21	0.30

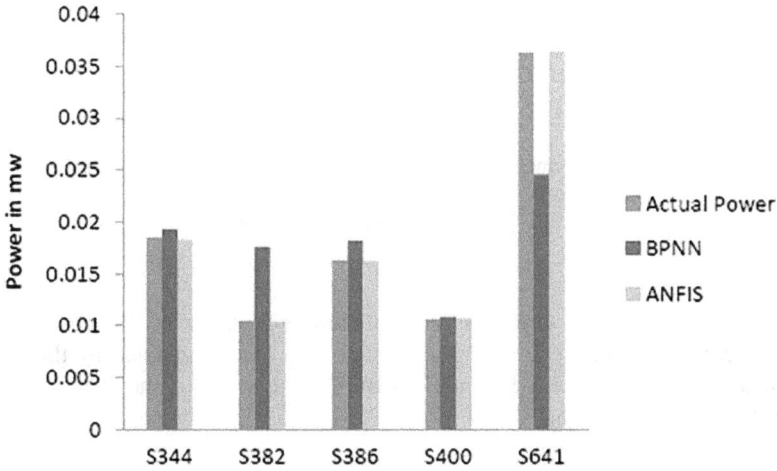

FIGURE 1.5 Comparison of actual power and power estimated by BPNN and ANFIS.

$$R = \frac{\sum_{i=1}^{N}\left(Y_I^O - Y^O\right)\left(Y_I^C - Y^C\right)}{\sqrt{\sum_{I=1}^{N}\left(Y_I^O - Y^O\right)^2 \sum_{I=1}^{N}\left(Y_I^C - Y^C\right)^2}} \qquad (1.3)$$

Where Y_I^O is the observed value and Y_I^C is the calculated value; similarly, Y^O is the mean of the observed value and Y^C is the mean of the calculated value. RMSE is usually used as a measure to find the difference between values predicted by the model and measured original values. RMSE can be used to indicate model accuracy

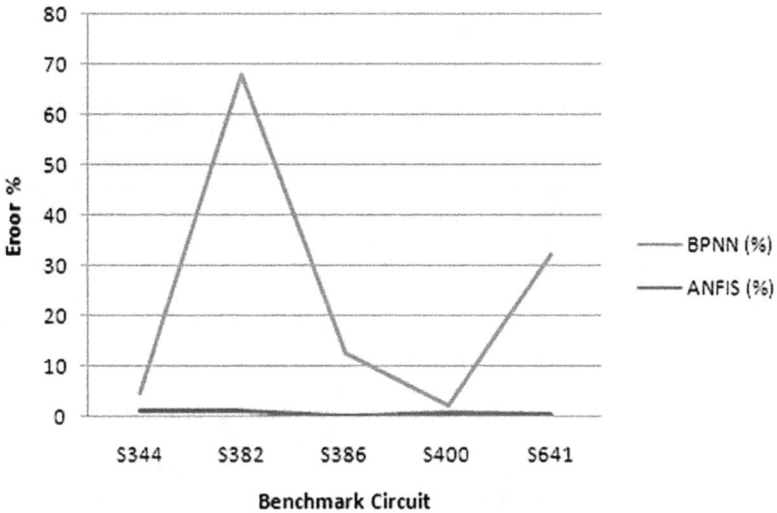

FIGURE 1.6 ANFIS and BPNN error percentage comparison.

TABLE 1.11
Statistical Analysis Test Comparison of BPNN and ANFIS

Parameter	BPNN	ANFIS
RMSE	0.0004499	0.0002075
R	0.84696	0.99961

or precision and the value of RMSE should be low or close to zero. The R determines a linear correlation between measured values and values simulated by the model, whose optimal value is 1. RMSE and R for BPNN and ANFIS is shown in Table 1.11; from the table we infer that RMSE of ANFIS is less when compared to BPNN and R value of ANFIS is very close to 1.

1.6 CONCLUSION

Since power estimation for CMOS VLSI circuits using SPICE simulation is time consuming and it is desirable to do power estimation at a high level of abstraction with less design time and cost, we have proposed an approach using statistical techniques such as ANFIS and BPNN. Conventional techniques like SPICE simulation are based on the assumptions of predefined empirical equations that depend on arbitrary parameters, so power estimation using ANFIS and BPNN can be an alternative solution to SPICE simulation. The results of BPNN and ANFIS are highly accurate, and unlike other systems, details of the circuit structure and interconnections are not required. ANFIS gives a better result in terms of testing error that varies from 0% to 0.86% when compared to BPNN. It also has a low RMSE value of 0.0002075 and a very high R value of 0.99961.

REFERENCES

1. Kozhaya, J. N., & Najm, F. N. (2001). Power estimation for large sequential circuits. IEEE Transactions on Very Large Scale Integration Systems, 9(2), 400–407. http://dx.doi.org/10.1109/ICCAD.1997.643581.
2. Akdagli, A., Kayabasi, A., & Develi, I. (2015). Computing resonant frequency of C-shaped compact microstrip antennas by using ANFIS. International Journal of Electronics, 102(3). http://dx.doi.org/10.1080/00207217.2014.897379.
3. Al-Shammari, E. T., Amirmojahedi, M., Shamshirband, S., Petković, D., Pavlović, N. T., & Bonakdari, H. (2015). Estimation of wind turbine wake effect by adaptive neuro-fuzzy approach. Flow Measurement and Instrumentation, 45, 1–6. http://dx.doi.org/10.1016/j.flowmeasinst.2015.04.002.
4. Bhanja, S., & Ranganathan, N. (2003). Switching activity estimation of VLSI circuits using Bayesian networks. IEEE Transactions on Very Large Scale Integration Systems, 11(4), 558–567. http://dx.doi.org/10.1109/TVLSI.2003.816144.
5. Burch, R., Najm, F. N., Yang, P., & Trick, T. N. (1993). A Monte Carlo approach for power estimation. IEEE Transactions on Very Large Scale Integration Systems, 2(1), 63–71. http://dx.doi.org/10.1109/92.219908.

6. Saxena, V., Najm, F. N., & Hajj, I. N. (1997). Monte-Carlo approach for power estimation in sequential circuits. European Design and Test Conference, 17–20 March (ED&TC 97), 416–420. http://dx.doi.org/10.1109/EDTC.1997.582393.
7. Chiu, S. (1996). Method and software for extracting fuzzy classification rules by subtractive clustering, in Proceedings of Fuzzy Information Processing Society NAFIPS Biennial Conference North American, pp.461–465. http://dx.doi.org/10.1109/NAFIPS.1996.534778.
8. Klein, F., Leao, R., Santos, G. A., & Azevedo, R. (2009). A multi-model engine for high-level power estimation accuracy optimization. IEEE Transactions on Very Large Scale Integration Systems, 17(5), 660–673. http://dx.doi.org/10.1109/TVLSI.2009.2013627.
9. Buyuksahin, K. M., & Najm, F. N. (2006). Early power estimation for VLSI circuits. IEEE Transactions on Computer Aided Design of Integrated Circuits and Systems, 24(7), 1076–1088. http://dx.doi.org/10.1109/TCAD.2005.850904.
10. Karimi, G., Banitalebi, R., & Sedaghat, S. B. (2013). Simulation of SiGe:C HBTs using neural network and adaptive Neuro-fuzzy inference system for RF applications. International Journal of Electronics, 100(7). http://dx.doi.org/10.1080/00207217.2012.727353.
11. Guney, K., & Sarikaya, N. (2007). Adaptive neuro-fuzzy inference system for computing the resonant frequency of electrically thin and thick rectangular micro strip antennas. International Journal of Electronics, 94(9), 007. http://dx.doi.org/10.1080/00207210701526317.
12. Güler, İ., & Übeyli, E. D. (2005). Adaptive neuro-fuzzy inference system for gap discontinuities in coplanar waveguides. International Journal of Electronics, 92(3). http://dx.doi.org/10.1080/00207210512331337668.
13. Ramesh, J., Vanathi, P.T., & Gunavathi, K. (2008). Fault classification in phase-locked loop using back propagation neural network. ETRI Journal, 30(4), 546–554. http://dx.doi.org/10.4218/etrij.08.0108.0133.
14. Karthika, B. S., & Dek, P. C. (2015). Prediction of air temperature by hybridized model (Wavelet- ANFIS) using wavelet decomposed data. Aquatic Procedia, 1155–1161. http://dx.doi.org/10.1016/j.aqpro.2015.02.147.
15. Hou, L., Zheng, L., & Wu, W. (2006). Neural Network Based VLSI Power Estimation. IEEE International Conference on Solid State and Integrated Circuit Technology, 2006, pp.1919–1921. http://dx.doi.org/10.1109/ICSICT.2006.306506.
16. Mohammadi, K., Shamshirband, S., Yee, P. L., Petković, D., Zamani, M., & Ch, S. (2015). Predicting the wind power density based upon extreme learning machine. Energy, 86, 232–239. http://dx.doi.org/10.1016/j.energy.2015.03.111.
17. Murugavel, A. K., Ranganathan, N., Chandramouli, R., & Chavali, S. (2002). Least-square estimation of average power in digital CMOS circuits. IEEE Transactions on Very Large Scale Integration Systems, 10(1), 55–58. http://dx.doi.org/10.1109/92.988730.
18. Nikolić, V., Motamedi, S., Shamshirband, S., Petković, D., Ch, S., & Arif, M. (2015). Extreme learning machine approach for sensor less wind speed estimation. Mechatronics, 34, 78–83. http://dx.doi.org/10.1016/j.mechatronics.2015.04.007.
19. Nikolić, V., Petković, D., Por, L. Y., Shamshirband, S., Zamani, M., Ćojbašić, Ž., & Motamedi, S. (2016). Potential of neuro-fuzzy methodology to estimate noise level of wind turbines. Mechanical Systems and Signal Processing, 66–78, 715–722. http://dx.doi.org/10.1016/j.ymssp.2015.03.013.
20. Nikolić, V., Petković, D., Shamshirband, S., & Ćojbašić, Ž. (2015). Adaptive neuro-fuzzy estimation of diffuser effects on wind turbine performance. Energy, 89, 324–333. http://dx.doi.org/10.1016/j.energy.2015.05.126.

21. Nikolić, V., Shamshirband, S., Petković, D., Mohammadi, K., Ćojbašić, Ž., Altameem, T. A., & Gani, A. (2015). Wind wake influence estimation on energy production of wind farm by adaptive neuro-fuzzy methodology. Energy, 80, 361–372. http://dx.doi.org/10.1016/j.energy.2014.11.078.

22. Dalibor, P., Ćojbašić, Z., & Nikolić, V. (2013). Adaptive neuro-fuzzy approach for wind turbine power coefficient estimation. Renewable and Sustainable Energy Reviews, 28, 191–195. http://dx.doi.org/10.1016/j.rser.2013.07.049.

23. Petković, D., Pavlovic, N. T., Shamshirband, S., Kiah, M. L. M., Anuar, N. B., & Idris, M. Y. I. (2014). Adaptive neuro-fuzzy estimation of optimal lens system parameters. Optics and Lasers in Engineering, 55, 84–93. http://dx.doi.org/10.1016/j.optlaseng.2013.10.018.

24. Dalibor, P., Shamshirband, S., Pavlovi, N. T., Anuarc, N. B., & Mat Kiahc, M. L. (2014). Modulation transfer function estimation of optical lens system by adaptive neuro-fuzzy methodology. Optics and Spectroscopy, 117(1), 121–131. http://dx.doi.org/10.1134/S0030400X14070042.

25. Petković, D., Ćojbašić, Ž., Nikolić, V., Shamshirband, S., Kiah, M. L. M., Anuar, N. B., & Wahab, A. W. A. (2014). Adaptive neuro-fuzzy maximal power extraction of wind turbine with continuously variable transmission. Energy, 64, 868–874. http://dx.doi.org/10.1016/j.energy.2013.10.094.

26. Petković, D., Ab Hamid, S. H., Ćojbašić, Ž., & Pavlović, T. N. (2014). Adapting project management method and ANFIS strategy for variables selection and analyzing wind turbine wake effect. Natural Hazards, 74(2), 463–475. http://dx.doi.org/10.1007/s11069-014-1189-1.

27. Petković, D., Shamshirband, S., Ćojbašić, Z., Nikolić, V., Anuar, N. B., Sabri, A. Q. M., & Akib, S. (2014). Adaptive neurofuzzy estimation of building augmentation of wind turbine power. Computers & Fluids, 97, 188–194. http://dx.doi.org/10.1016/j.compfluid.2014.04.016.

28. Petković, D., Shamshirband, S., Anuar, N. B., Naji, S., Kiah, M. L. M., & Gani, A. (2015). Adaptive neuro-fuzzy evaluation of wind farm power production as function of wind speed and direction. Stochastic Environmental Research and Risk Assessment, 29(3), 793–802. http://dx.doi.org/10.1007/s00477-014-0901-8.

29. Petković, D., Shamshirband, S., Tong, C. W., & Al-Shammari, E. T. (2015). Generalized adaptive neuro-fuzzy based method for wind speed distribution prediction. Flow Measurement and Instrumentation, 43, 47–52. http://dx.doi.org/10.1016/j.flowmeasinst.2015.03.003.

30. Petković, D. (2015). Adaptive neuro-fuzzy approach for estimation of wind speed distribution. Electrical Power and Energy Systems, 73, 389–392. http://dx.doi.org/10.1016/j.ijepes.2015.05.039.

31. Ramanathan, P., Surendiran, B., & Vanathi, P.T. (2013). Power estimation of benchmark circuits using artificial neural networks. Pensee Journal, 75(9), 427–433. https://www.researchgate.net/publication/257939769_Power_Estimation_of_Benchmark_Circuits_using_Artificial_Neural_Networks.

32. Lorenzo, R., & Chaudhry, S. (2016). Review of circuit level leakage minimization techniques in CMOS VLSI circuits. IETE Technical Review, April 2016, 1–23. http://dx.doi.org/10.1080/02564602.2016.1162116.

33. Shamshirband, S., Petković, D., Nikolic, Z. C. V., Anuar, N. B., Mohd Shuib, N. L., Mat Kiah, M. L., & Akib, S. (2014). Adaptive neuro-fuzzy optimization of wind farm project net profit. Energy Conversion and Management, 80, April 2014, 229–237. http://dx.doi.org/10.1016/j.enconman.2014.01.038.

34. Shamshirband, S., Petković, D., Anuar, N. B., & Gani, A. (2014). Adaptive neuro-fuzzy generalization of wind turbine wake added turbulence models. Renewable and Sustainable Energy Reviews, 36, 270–276. http://dx.doi.org/10.1016/j.rser.2014.04.064.

35. Shamshirband , S., Petković, D., Hashim, R., & Motamedi, S. (2014). Adaptive neuro-fuzzy methodology for noise assessment of wind turbine. PLoS One, 9(7), 1–9. http://dx.doi.org/10.1371/journal.pone.0103414.

36. Shamshirband, S., Petković, D., Hashim, R., Motamedi, S., & Anuar, N. B. (2014). An appraisal of wind turbine wake models by adaptive neuro-fuzzy methodology. Electrical Power and Energy Systems, 63, 618–624. http://dx.doi.org/10.1016/j.ijepes.2014.06.022.

37. Shamshirband, S., Petković, D., Tong, C. W., & Al-Shammari, E. T. (2015). Trend detection of wind speed probability distribution by adaptive neuro-fuzzy methodology. Flow Measurement and Instrumentation, 45, 43–48. http://dx.doi.org/10.1016/j.flowmeasinst.2015.04.007.

38. Govindaraj, V, & Ramesh, J. (2018). Adaptive neuro fuzzy inference system-based power estimation method for CMOS VLSI circuits. International Journal of Electronics, 105(3), 398–411 (Taylor and Francis, ISSN 0020-7217). http://dx.doi.org/10.1080/00207217.2017.1357763.

39. Harris, C. J. (1994). Advances in intelligent control. CRC Press. ISBN 9780748400669.

2 Awareness Alert and Information Analysis in Social Media Networking Using Usage Analysis and Negotiable Approach

Alamelu Muthukrishnan
Kumaraguru College of Technology

CONTENTS

2.1 INTRODUCTION

Today Internet communication is the smart connectivity link between different people. In olden days, communication was carried out with letters, post cards, telegrams, and land telephones. People from different place communicated through any one of many types of communication. In cases of message delivery or emergency information sharing, data can be shared either with the postal card delivery or telegram message delivery. In these cases, accessing the communication particulars sometimes got delayed because of unconditional weather conditions or unpredictable situations. In this scenario, people may be adversely affected by the delay of receiving valuable information.

To accomplish quick communication, the technology can be improved with the digital and information technology techniques. Using Internet and intranet technologies, information can be accessed from anywhere at any point of time. Initially, the widespread launch of the Internet has spread communications from people from remote resources. Now the Internet has spread and diversified and can be accessed with social media networks either using web technologies or mobile applications. People are accessing quickly relevant messages with social networks like Facebook, Twitter, WhatsApp, Instagram, Skype, Google+, and so on.

These social networks may play a vital role in every human being's life to tackle their day-to-day life information updates. In particular, the social networks can have both advantages and disadvantages. The advantages include the diversity of people's information sharing: quick posting of messages, likes, and updates of posts as well as delivering products for business landlords. People have the fast communication and accessibility of information from any sort of networks. The scenario also has disadvantages, like misuse of data, privacy, posting of the other's user account data, and information hacking. These can negatively affect the social networking user's life and create many unsolvable problems. The major area affected by this is the misuse of the person's login identification, such as one person's login credentials being used by the another person and misusing a person's data information.

There are many possibilities to hack data in a social media network. The user should be aware about privacy concerns when posting any personal information in the social network. Some concerns are the misuse of posted photos and hacking personal information in the social networks. Hackers can use hacking techniques and tools to hack any user data. Such hacked information or data is misused by the hackers. Due to these disadvantages, it can be difficult to solve the major issues of the resulting social problems.

2.1.1 LITERATURE SURVEY

Zhuozhao Li et al (2018) [1] has proposed the concept called the Disney Information Station (DIS boards), a discussion forum for the Disney-related travel for people who visit Disney properties frequently. The DIS system was developed to reduce the search time of Disney-relevant information such as resorts, dining, and other people's comments to the other visitors. With respect to the usage of DIS, teenagers and adult users have been categorized by the system as clusters of report, fact, and discussion. In the system thread categories are created and communicated for the discussion forum. The cluster analysis called ego network has been created for trip reports, restaurants, and adventures. By the DIS system, the Disney social network users can quickly find relevant information quickly.

Ming Cheung et al (2018) [2] have researched the search engine social network for the west and east content-sharing mechanism. In this mechanism, the authors have referenced eight million people's usage with eight different social networks. In this work the major analysis started with user-shared images and the friend's followers. The search engine process began by collecting images from multiple social networks and the images are labeled and discovered with a concept called convolutional neural network. By connecting more than two or three social networks the connectivity and discoverability can be improved for the followers in the social networks.

Rahil Sharma et al (2017) [3] have proposed an algorithm called novel hybrid parallel algorithm to categorize the community of the groups in the social networks. As social networks have developed, diverse parallel algorithms a parallel algorithm with synthetic graphs has been used to categorize the analysis. In this approach, in order to cluster the community groups, the multilevel multicore (MCML) community detection algorithm (shared memory parallel implementation) has been used to divide the groups into clusters. In this the levels are categorized as network partitioning, renumber vertices and image partitions, and MCML. As with the different partitioning, the good scalability and proper quality of community searching can occur in the social communities.

Xue Yung et al (2016) [4] has introduced the protocol called secure and fine-grained privacy-preserving matching (SFPM) protocol for secure mobile communications. The protocol was developed to enhance the security of mobile communications. In SFPM, two methods are used to describe the matches between the two mobiles, such as the cosine similarity check and weighted L1 norm matching. For the first one, a similar set of objects are identified and measured with the cosine similarity verification. In the second, weighted L1 matching was preferred to match the weightages of the object during the communication transmission. As with the usage of the SFPM protocol, only authenticated information was shared safely in Android mobile applications. The SFPM protocol was tested with two types of Android mobile applications and one PC. In this approach the user can secure information in the Androids using SFPM protocols.

Hongjian Wang et al (2019) [5] proposed that crime rate analysis can be executed in different research. In this chapter, the authors proposed the term Point of Interest (POI) data in the area of Chicago. A binomial regression model and a geographically weighted regression model were developed to monitor the features of theft, criminal damage, burglary, and motor vehicle theft. Using the POI approach the crime analysis can be quickly analyzed and rectified.

Daniel Zhang et al (2019) [6] have proposed the scalable and robust truth discovery (SRTD) approach to study identical misinformation spread on the social media. The chapter focuses on Twitter analysis based on the truth fullness scores. The

SRTD can be analyzed with data sparsity analysis and data fusion techniques. Three main observations are monitored and deployed to improve the SRTD algorithm: very lagging independent verification during forwarding the messages, forwarding the false claims, and false consideration about the previous claims. Using this approach the misinformation spread in the social networks Twitter, Facebook, and Instagram were analyzed.

Alamelu Muthukrishnan et al (2017) [7] proposed a method for the most people usage (MPU) Internet of Things structure for the health care and social networking systems. The collective healthcare-related queries are collected from the end user and analyzed with the expert members. The most unsolved queries are input to the proposed Health Social Cloud Center (HSCC) system. The internal system will make the analysis of the risk analysis and expert member discussion with the expert analysis databases. The finalized solution is then sent to the requested client and the customer can make the final decisions.

2.2 USAGE ANALYSIS AND NEGOTIABLE (UAN) APPROACH

In general, many societal issues may occur through misuses of social media networks. To reduce misuse of social networks and to create awareness in teenagers and adults, the proposed usage analysis and negotiable (UAN) approach will propose user view analysis with the user interested test rate (UITR) analysis algorithm approaches. The proposed UAN system may subdivided into a number of submodules categorized by age categorization usage analysis, user view analysis, and UITR analysis algorithm and feedback generation.

In this analysis the end user can use various social networking applications and share their full thoughts in the social networking system. Initially the system will categorize end-user usage with the age categorizations and, based on the category of ages 18–35, it next puts forward to the user view analysis. In this case, the user information sharing can be categorized with the likes and dislikes, content sharing, and comment-posting categories. With respect to the user usage analysis, the UITR algorithm will rate the user usage of the networks to the end users. With this the end users, especially the age group from 18 to 35, can come to the valuable decisions of using the social networks in a good way. The detailed processing UAN approach is depicted in Figure 2.1.

The initial flow of processing of UAN can start from the diverse set of end customers. The customers can choose the social networks. They will use the diverse version of the social network applications such as Facebook, WhatsApp, Instagram, Twitter, Google +, Vimeo, and You Tube.

As the initial step, the age categorization can be carried out with the set of age groupers. The first step can be started from the age categorization. The age categorization can be started with the adult age categorization analysis, in this case the age categorization can be carried out as the group of 18 to 35 age groupers. The age categorization analysis can then be carried out with the categories of 18 to 25, 26 to 30, and 31 to 35.

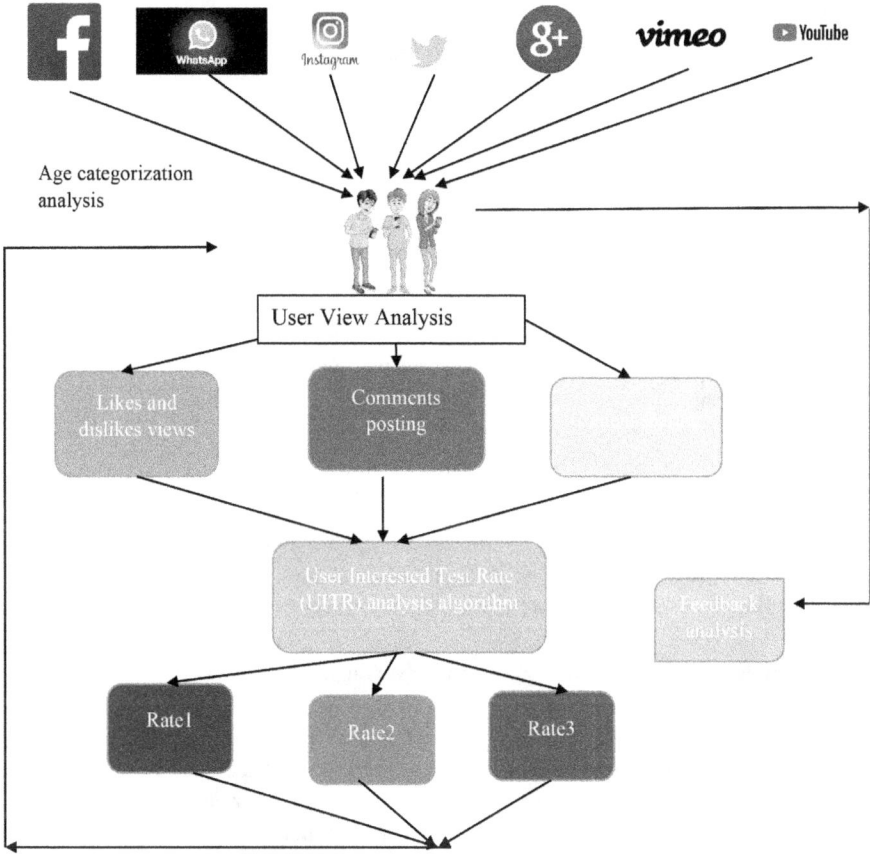

FIGURE 2.1 Usage analysis and negotiable approach.

2.3 AGE CATEGORIZATION USAGE ANALYSIS

The proposed UAN system will start with the initial age categorization with the category of the age groupers from 18 to 35. This can be divided with the categorization from 18 to 25, 26 to 30, and 31 to 35 groupers. As per the adult age categorization, the rule states for beginners, medium agers, and large age groups.

2.3.1 RULES DEFINED FOR 18–25 AGE CATEGORIZATION

This age group is defined as the beginners of the UAN system. In this case, the system will recognize the beginners with the initiation rules for the identification. That is:

1. Beginner has to verify their details with the DOB profile verification.
2. Beginner Entry profile verification = if (DOB = 18 or DOB <= 25) then verify Government ID proof.

3. Else
4. Beginner Entry profile verification = Restricted category
5. The DOB profile verification can linked with any one of the government-defined proof of voter ID or Aadhaar number.
6. If the age categorization does not match with the beginner identification it can be strictly monitored with the policy awareness identification rules.
7. If matching has not been found then the unmatched beginners are put on the restricted categorization.

According to the beginners awareness policies, unmatched beginners are instructed to avoid preregistration and they are put on a waiting list until they meet the eligibility criteria.

2.3.2   RULES DEFINED FOR 26–30 AGE CATEGORIZATION

The second category is defined for the middle age group from 26 to 30 years of age. Those grouped in this category have more awareness than the young age group. The 26–30 age group starts with the following rules:

1. Initially verified with the DOB profile verification.
2. With respect to the DOB, any one of the government-defined documents can be verified (as similar to beginners policy)
3. The middle age group is verified with the usage of the social media networks, for example, a count of social networks for a person.
4. Middle age category verification = (Social network account count + Active usage of the social network)/(Person choice of using the social account).
5. If a person has more uncounted social networking accounts (more than 7), the advantages and disadvantages of using the specific social networks can be sent as a report to the end users.

2.3.3   RULES DEFINED FOR 31–35 AGE CATEGORIZATION

This rule can be implemented for the 31 to 35 age group. At this age end users mostly have more awareness about the social networks. If they have less awareness based on their usage, the general report is sent to the end users.

The social network questionnaire survey score card data has been collected for the 31 to 35 age group.

Adult age category verification = Social network survey analysis

The rating has been fixed based upon the score card attained by users in this age group. If the user has a low score about social networking issues then the report will upgrade them with the advantages and disadvantages of social networks.

Eligibility of adult age grouper =
If the score card = 8.6/10 to 10/10 -> High awareness
If the score card = 6/10 to 8.5/10 -> medium awareness
If the score card = less than 5.9/10 -> Low awareness

The above age categorical analysis has been tested before using the UAN system. After the age categorization verification, the system will next evaluate the user view analysis

2.4   USER VIEW ANALYSIS

The user view analysis can be generated based on the end user's usage of the social networking systems, such as Facebook, WhatsApp, Instagram, Google +, and so on. Based on the usage the users can be categorized by:

- Likes and dislikes views
- Comments posting
- Content sharing

Figure 2.2 represents the user view analysis of a sample image that has been posted on social media and has been liked and disliked by the users with comments posting and content sharing.

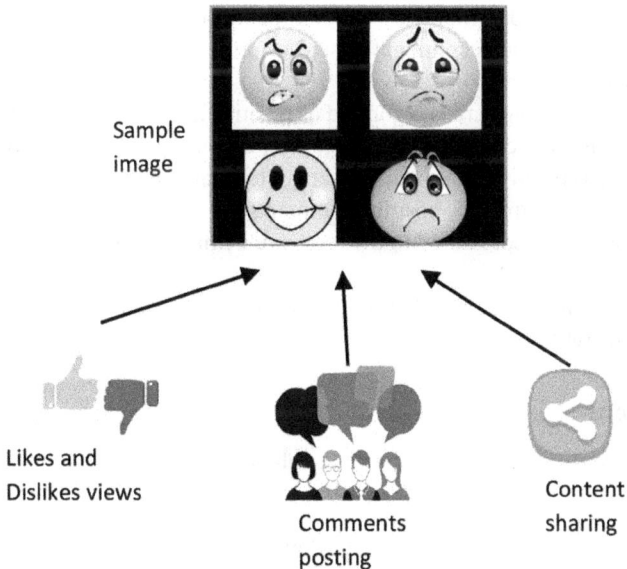

FIGURE 2.2   User view analysis.

2.4.1 Likes and Dislikes Views

Likes and dislike views are analyzed with the count of likes and dislikes that the user posted to the content. Content domain can be chosen based upon the likes and dislike views.

Choose the Domain of the content posted in the social network.

Select the likes and dislikes view count based on the content posting with respect to the day.

If the content is posted within 2 days then it to be matched with the number of count of likes. If the count has increased it can be mapped with the days.

Number of likes and dislikes	<= 15 within two days comes under Low priority
Number of likes and dislikes	>= 16 to 75 within seven days comes under the medium priority
Number of likes and dislikes	>= 76 up to the unlimited count of days then it comes under the highest priority.

Based on the likes and dislikes usage category the user Likes and Dislikes report has been generated and defined in different categories.

2.4.2 Comments Posting Category

The comment posting can be defined with the category of positive and negative comments. With respect to the categorization of comments the content posted in the social network can be identified with the following steps:

1. Comments are identified with the comments posted timing.
2. Comment posting = comment posted timing + the positive comments (or) negative comments
3. Comment postings are categorized as positive and negative comments.
4. If the comment comes under the generic formal quotable notations, it can be taken as positive comments . User quoted statements or comments in this category are defined as positive comments.
5. The comments not related to the topic are defined as irrelevant comments.

Based on the positive and negative types of comments the posting timing is noted and sent to the rating analysis.

2.4.3 Content Sharing

The content sharing can be based on user choice and their willingness to post the content. The content-sharing process follows:

1. User posting contents can be viewed with respect to the user view choices.
2. Content sharing can be chosen based on the likes viewed by that content and the number of users.

3. Sharing to other users depends upon the set of friends or the specific friend category.
4. If the set of friend's category has the highest count then the user content sharing can be matched with the type of domain shared to the user.
5. List of Friends sharing category = Number of friends list + most preferred
6. Domain shared
7. If the single user content has been shared then it has been rated with the types of content sharing with the domain of the content.
8. Single user sharing category = single friend + specific content shared

With the conclusion of the three categories of user view analysis (likes and dislikes views, comments posting, and content sharing) the final disclosure report is sent to the rating analysis system for user interested rate analysis.

2.5   USER INTERESTED TEST RATE (UITR) ANALYSIS ALGORITHM

The UITR algorithm will test the analysis of the data based on the user likes and dislikes views, comments posting, and content sharing of the end users. The rating can be based on the user interested usage of the social networks. Figure 2.3 represents the UITR analysis algorithm with the three processing internal execution.

2.5.1   RATING STATEMENT ANALYSIS

Rate analysis can be based upon the review analysis of the expert member's suggestions and discussions with respect to the report of the user view analysis.

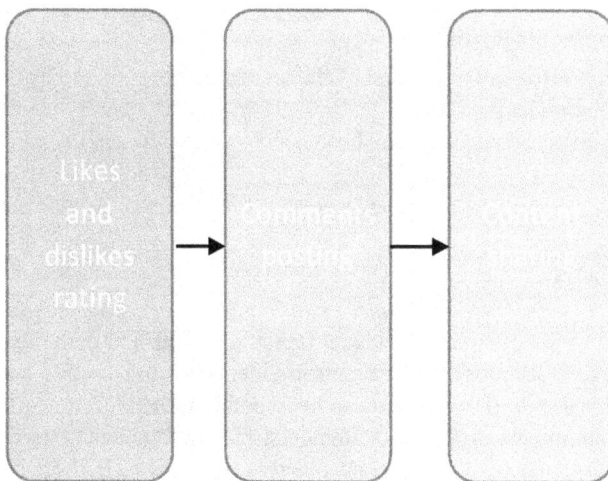

FIGURE 2.3   User interested test rate (UITR) analysis.

2.5.1.1   Like and Dislikes Rating Algorithm

1. If the likes and dislikes are viewed with the category of
   - Number of Likes <=15 for low priority,
   - Number of Likes >=16 to 75 for medium priority
   - Number of Likes >=76 with the high priority
2. The rating can be analyzed with respect to day, hour of posting and based on the type of domain.
3. If the category is of <=15 for the low priority then the analysis can be noted with in two days of duration
4. Calculation of Likes & dislikes views

   $\sum$ Likes & dislikes views = $\sum$ Total posting likes and dislikes analysis duration

   Within two days/Total hours of posting (24 hours)

   For example,

   Scenario 1: if the likes and dislikes posting within 2 days are the count of 15 matched with 24 hours then the resulting output is of 15/24 = 0.625.

   Scenario 2: if the likes and dislikes posting within 2 days are the count of 35 matched with 24 hours then the resulting output is of 2/35 = 1.458

   Scenario 3: if the likes and dislikes posting within 2 days are the count of 80 matched with 24 hours then the resulting output is of 80/24= 3.333
5. The highest priority match has been identified and mapped with the domain posting.

   The domain can be of picture or content.
6. With respect to the domain of content a report has been generated about the content and sent to the user. The report will depicts if the picture is being shared with the friends list or friend to friend list.
7. By viewing the report analysis the user may have the knowledge of priority analysis about the domain sharing with their friends.

2.5.1.2   Comments Posting

The comment posting can be defined by the set of comments posted by the users with the sort selected topics.

Comment-posting category can be categorized based on the analysis of the domain.

Domain analyses are categorized as general topics or societal relevant topics.

Domain analysis = General topics or Societal topics/levels - > 0, 1, 2

If the general topics or societal relevant topics are identified then the positive and negative categorization of the comments are identified and divided with the levels of zero, one, and two. If level zero has been defined, then it is to be defined with the positive comments of formal statements. If level one has been defined, then it is to be defined with the moderate positive comments. If level two has been defined, then it is to be defined with the negative comments such as non-formal statements.

Based on the domain levels of rating a report is generated and sent for the customer's analysis.

2.5.1.3   Content Sharing

Content sharing can be defined with the types of content to be shared to the friends or the followers.

The content sharing can be chosen with the type of domain to be shared.

1. Choose any one category, either List of Friends sharing category or Single user sharing category
2. The chosen category can be mapped with the type of content shared. The content shared in the highest count is identified and mapped to the user.
3. Content sharing = List of Friends sharing category or Single user sharing category + type of content shared with highest priority

2.6   FEEDBACK ANALYSIS

Feedback analysis is executed from the end users with the user reports. The user has the view of the final report analysis sent from the (UITR) analysis algorithm. With respect to the user usage analysis the system will produce the analysis report for the end user, and the user can also view the analysis with their social network usage. Finally, the user can make a conclusion about using the social networks with given their advantages and disadvantages.

2.7   RESULT AND DISCUSSIONS

The proposed UAN approach is one approach to providing an awareness alert to youngsters and teenagers about using social networks. Many teens and adults are affected by issues of improper use of social networks. The proposed UAN system will analyze the issues with respect to age criteria and will provide a rating report based on age criteria and usage of the social networks. With reference to the existing systems, the proposed approach can take the user likes and dislikes reference, comment posting, and content-sharing categories. Based on these categories the UTTR algorithm has been defined to rate the user utilization of the social networks. Table 2.1 depicts the comparative study analysis of the proposed UAN approach with the existing methods.

Compared to existing study approaches, the proposed UAN study analysis can conclude the existing properties and improve the enhancement of services with the discussed properties. Table 2.2 depicts the improved analysis properties from the existing systems with the proposed system.

Compared to the existing study analysis methods, the proposed UAN approach can improve the social network analysis with the properties of the priority-based rating with respect to the user utilization of the social networks. Based on this the, teens and young adults can decide to use the social media networks with their advantages and disadvantages.

TABLE 2.1

Existing Study Analysis

S. No	Author	Existing Methods/ Technology Used	Application Used	Advantages	Characteristics Defined
1	Zhuozhao Li et al (2018)	DIS boards	Disney-related travel information (resorts, dining)	People visiting Disney can use this dashboard	Searching
2	Ming Cheung et al (2018)	West and east content-sharing mechanism.	Content sharing e.g., follower, images	Different people from different resources can use it	Information sharing
3	Rahil Sharma et al (2017)	Novel hybrid parallel algorithm	Identification of social community sharing	Easy way to track the user community in the social networks.	Availability
4	Xue Yung et al (2016)	SFPM mobile communication	Enhance the security of mobile communication	Secure Mobile app communications	Security
5	Kun Kuang et al (2018) [8]	Propensity Score Matching (PSM) based method	To promote information in the social media networks	Used for the social media promotions	Information sharing

TABLE 2.2

UAN Approach Analysis

S. No	UAN Approach	Characteristics Proposed for End Users	Existing Analysis Approach Analyzed	Characteristics Included
1.	User view analysis		• West and east content-sharing mechanism	Information sharing
	• Likes and Dislikes views	Availability		
	• Comments posting	Security	• DIS boards	Searching
	• Content sharing		• Novel hybrid parallel algorithm	Availability
2	User Interested Test Rate (UITR) analysis algorithm	Information Sharing Priority-based rating		
3	Age Categorization analysis	End-user social media analysis with the age of 18 to 35	• SFPM mobile communication	Security
			• Propensity Score Matching (PSM) based method	Information sharing

2.8 CONCLUSION

The proposed User Analysis Negotiable (UAN) approach can provide solution and awareness about the social media usage for the teenagers and adults. The proposed approach can provide the solutions with the methods of age categorization analysis, user view analysis, and the UITR analysis algorithm. By studying the social analysis algorithms, users between the ages of 18 and 35 have clarity about social media network usage with its advantages and disadvantages. If they are less aware about social media, based on their utilization of the social networks, the proposed system UAN will provide the usage feedback analysis to the end users. In future, the work will be enhanced based on the connectivity of mobile application alert awareness messages by using deep learning algorithms and its techniques.

REFERENCES

1. Zhuozhao Li, Harrison Chandler, Haiying Shen (2018), "Analysis of Knowledge Sharing Activities on a Social Network Incorporated Discussion Forum: A Case Study of DISboards", IEEE transactions on big data, vol. 4, no. 4, pp. 432–446.
2. Ming Cheung, James She, Ning Wang (2018), "Characterizing User Connections in Social Media through User-Shared Images", IEEE transactions on big data, vol. 4, no. 4, pp. 447–458.
3. Rahil Sharma, Suely Oliveira (2017), "Community Detection Algorithm for Big Social Networks Using Hybrid Architecture", Big data research, vol. 10, pp. 44–52.
4. Xue Yung, Rongxing Lu, Hongbin Liang, Xiaohu Tang (2016), "SFPM: A Secure and Fine Grained Privacy Preserving Matching Protocol for Mobile Social Networking", Big data research, vol. 3, pp. 2–9 (Elsevier).
5. Hongjian Wang, Huaxiu Yao, Daniel Kifer, Corina Graif, Zhenhui Li (2019), "Non-Stationary Model for Crime Rate Inference Using Modern Urban Data", IEEE transactions on big data, vol. 5, no. 2, pp.180–194.
6. Daniel Zhang, Dong Wang, Nathan Vance, Yang Zhang, Steven Mike (2019), "On Scalable and Robust Truth Discovery in Big Data Social Media Sensing Applications", IEEE transactions on big data, vol. 5, no. 2, pp. 196–206.
7. Alamelu Muthukrishnan, Ramalatha Marimuthu (2017), "A Survey on Healthcare and Social Network Collaborative Service Utilization Using Internet of Things", Journal of advanced research in dynamical and control systems, vol. 9, pp. 1010–1030.
8. Kun Kuang, Meng Jiang, Peng Cui, Hengliang Luo, and Shiqiang Yang (2018), "Effective Promotional Strategies Selection in Social Media: A Data-Driven Approach", IEEE transactions on big data, vol. 4, no. 4, pp. 487–501.

3 Object Detection and Tracking in Video Using Deep Learning Techniques: A Review

Dhivya Praba Ramasamy and Kavitha Kanagaraj
Kumaraguru College of Technology

CONTENTS

3.1   INTRODUCTION

Computer vision is a broad area that describes how a machine is able to recognize data available in images or scenes. It replicates what human intelligence can do with the human visual system [1–3]. The multi-level data from the real world can be extracted using the following sequence of steps: capturing, processing, analyzing and understanding [4–7]. Object tracking [8, 9] has the following processes: object representation, object detection, and object tracking.

This chapter will give an idea of object tracking methods and their field of applications [10]. It will provide fundamental concepts to develop an object tracker. Different sources of image or video can be used, such as surveillance video, outlooks from multiple cameras, and medical scanners. The system can be developed by various theories and models.

In this chapter, Section 3.1 gives introduction. Section 3.2 focuses on challenges in visual tracking. Section 3.3 contains a detailed study on several fundamentals of object tracking methods. Section 3.4 explores a detailed study on feature extraction methods. Section 3.5 discusses various object classification methods. Section 3.6 focuses on various object tracking methods. Section 3.7 provides an introduction to machine and deep learning techniques. Section 3.8 discusses the results. Sections 9 and 10 contain the conclusion, future scope, and references.

3.2   CHALLENGES IN VIDEO TRACKING

Moving objects (single or many) viewed by a camera over a time period can be followed using video tracking. One major issue is object clutter. Often the object region of interest is similar to its background or sometimes it may be hidden by another object present in the scene. The presence of an object in the scene can cause problems for the following reasons (Figure 3.1 [11]):

- Object position: the presence of an intended object in a video frame may vary from one frame to another frame and also vary its projection on a video frame plane.
- Ambient illumination: the ambient light on the object of interest can change in intensity, direction, and color in a video plane.
- Noise: the video may contain some amount of noise during signal acquisition, based on the quality of the sensor during signal acquisition.
- Occlusions: a moving object in a scene may be hidden by another object present in the same scene; the object cannot be tracked even if it is present in the scene.

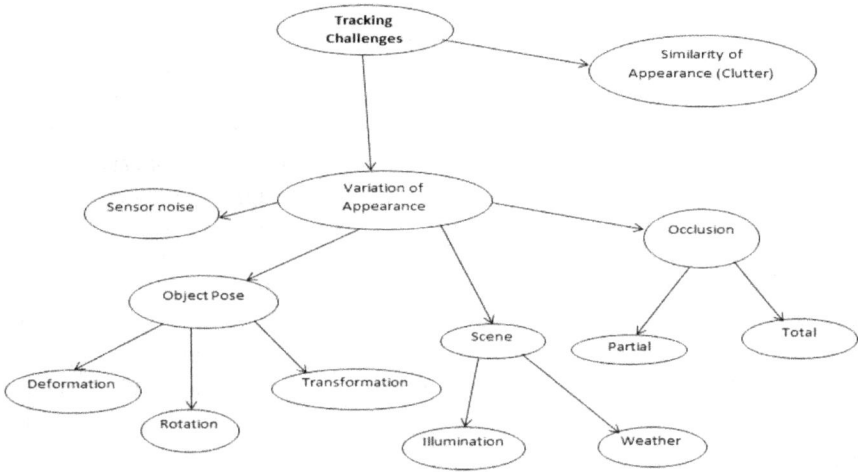

FIGURE 3.1   Challenges in visual tracking.

Some of the assumptions made during the tracking process are:

- Object motion is even.
- No sudden appearance variations take place.

Novel tracking methods can handle the following problems very easily: leaving objects out from scenes and drifting. The following parameters are necessary to build a good complete tracking model:

   i. Robustness

It is able to track the object even under difficult conditions. The difficult conditions are mainly caused by clutter background, change in the lighting conditions, blockages, or complicated object motion.

   ii. Adaptivity

In addition to changes in the environment, the object itself can undergo changes, which requires a good adaptation technique for the tracking model.

   iii. Real-time processing

Live videos requires a high-speed processing methods to track the object. The performance of the algorithms mainly depends on the motion of the object. The speed of the algorithm has to be 15 frames per second to maintain a good quality output video.

3.3   FUNDAMENTALS OF OBJECT TRACKING

Object tracking [8, 9] has the following processes: object representation, object detection, and object tracking.

3.3.1 OBJECT REPRESENTATION

Some applications of object tracking are tracing particular people in a video frame for video surveillance, tracking land-dwelling objects, or using satellite data for astral studies. Object selection depends on the application. If the application is traffic surveillance, then the object of interest may be a human, building, or car. For satellite applications the objects may be planets, whereas for gaming the object of interest may the human face. Figure 3.2 shows the region of interest in a video.

To keep the track of an object requires a pre-processing technique that converts the data into computer-understandable machine code. Shape and appearance form the basis, but extracted features are also used for object representation. Additional parameters that need to be included in object representation are application domain, persistence, and goals.

The representation determines the selection of best algorithm [10, 12]. In simple words object representation depends on shape representation and/or an appearance representation. Detailed descriptions of different shape representation methods are discussed below.

Representation of Object = Object Shape + Appearance

3.3.1.1 Shape Representation

The shape of an object can be represented using several methods, which also require some calculations for locating and tracing the object. It is essential to know the

FIGURE 3.2 Interested objects in video tracking (left) group of people, (right) face of single person.

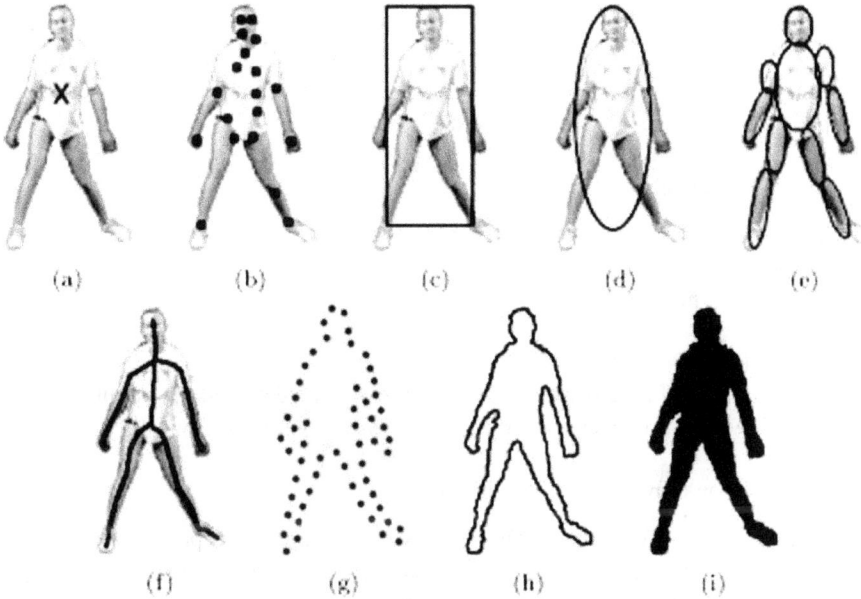

FIGURE 3.3 Shape representation techniques [13].

advantages and disadvantages of all the techniques because all methods are not appropriate for all applications. The general shape representations are discussed below and refer to Figure 3.3:

i. Points

Single (Figure 3.3(a)) or multiple points (Figure 3.3(b)) can be suitable for representing the objects.

These points can play a vital role if the object tracking is an image. The scenario considered can vary from single object tracking to many objects present in the scene so that interaction between objects can be obtained. These interactions may cause error. For simple and small objects, the point method is very suitable.

ii. Geometric shapes

Shape representation uses basic elementary shapes (Figures 3.3(c) and 3.3(d)). It is suitable for both fixed and moving objects. The nature of moving objects is very complex, so the most similar parts of the objects are included in the shape template.

iii. Silhouette and contour

This technique uses a skeleton or border (Figure 3.3(g), Figure 3.3(h)) for represent an object. Within that it uses another region (Figure 3.3(i)). It makes the representation easier and can be used to represent flexible and non-flexible objects. This model is able include any change wide range of object forms.

iv. Articulated shape models

Various parts of the object can be combined to form an articulated object. Various parts of the human being shown in Figure 3.3(e) are used during representation. Elliptical shapes are used to represent the object.

v. Skeletal model

A skeleton (Figure 3.3(f)) of the object can be extracted from the object outline. This is possible with medical axis transforms. This method is widely used in object recognition, but not when tracking objects.

3.3.1.2 Appearance Representation

There are number of methods available to express an object by its appearance. Some of the most widely used ways of representation are explained below [10].

i. Estimation of probability densities of object

Probability density functions (PDF) are used to express the probability of random variables [14]. With the help of a shape model, inner regions of an image can be recognized. For example, object appearance features can be determined using probability density functions. Some parameters are color or texture. The PDF can be either parametric (Gaussian distribution) or non-parametric (histograms).

ii. Templates

Templates can carry both appearance and spatial information. The template uses basic elementary shapes for representation. Templates are not suitable for challenging objects because they differ for different views. Object appearance parameters as well as object postures make the model more efficient. Templates create problems when the object features tend to change, for example when the lighting changes.

iii. Active appearance models

The shape of the object can be computed using the boundary of an object or object region. For each location the appearance of the object is modeled using color, texture, or gradient.

iv. Multi-view appearance models

Various outlooks of the object can be encoded using this method. Many methods are available. From the various views the model generates a subspace. This has been used in principal component analysis (PCA) and independent component analysis (ICA) [15].

3.3.2 OBJECT DETECTION

The fundamental procedure is to find the object that needs to be tracked in a video scene. Next, cluster the pixels of these objects. This involves the steps below. The

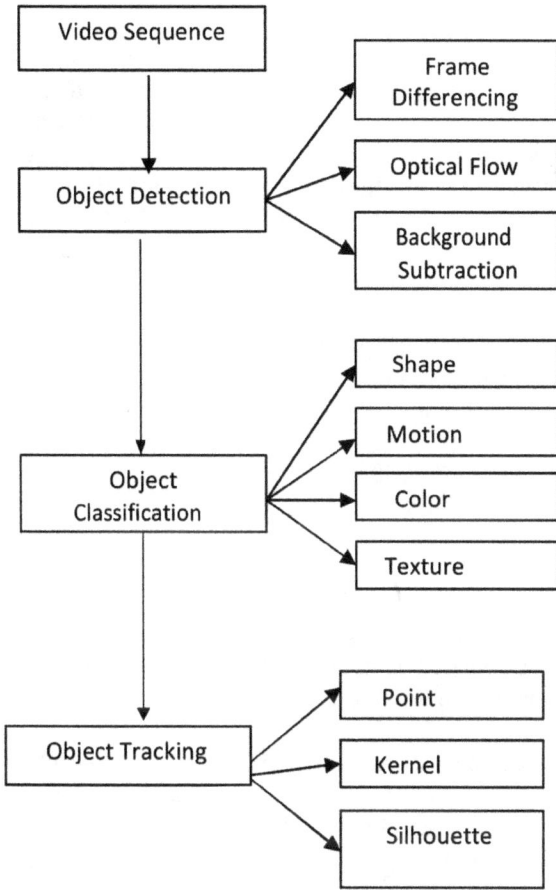

FIGURE 3.4 Fundamental steps in object tracking.

main data to be extracted is moving objects. Almost all the methods focus on object detection. The object detection in a video can be found when the object is entering into the video [10]. Object detection types are shown in Figure 3.4.

3.3.2.1 Frame Differencing

The target in a video sequence can be identified by subtracting the target from successive images or frames. For dynamic environments this method has strong adaptability. However, locating the moving object becomes very complicated due to the unavailability of complete outline information [16].

3.3.2.2 Optical Flow

Optical flow is used to cluster an image. Entire data can be extracted from the background. Real-time applications sometimes depend on the following parameters: noise sensitivity, anti-noise performance, etc.

3.3.2.3   Background Subtraction

Background modeling provides a basis model. The presence of a moving object in a video sequence can be identified by matching every frame with the reference model. Background subtraction uses different methods to find these objects. Even though the implementation is easy, it is more sensitive to surrounding factors. The total object data can be extracted only when the background is known. For real-time applications the static background model cannot produce good results. Some of the factors influencing the background changes are reflections, animated images, and indoor scenes. Static backgrounds face great difficulties with outdoor scenes. Background subtraction based motion detection or tracking systems need to hold in the following critical situations [9]: noisy image and lighting conditions, minor movements of non-static objects due to the wind, movements of objects, shadow regions and multiple objects in the scene. There are two approaches:

  i. Recursive algorithm [17, 18]
 ii. Non-recursive algorithm

3.3.3   OBJECT CLASSIFICATION METHOD

The object classification method assigns a class label based on the features extracted. Object features can be based on its size, color, structure, and motion. Features used for object classification are listed below:

  i. Edges: intense intensity variations are found near the object boundary. Edge detection techniques are applied to calculate these intensity variations. Edges are less sensitive to lighting conditions [18].
 ii. Motion: object classification methods produce good results for non-rigid objects [19].
iii. Color: many color spaces are available to store data from different frames.
 iv. Texture: texture is used to identify the target or object of interest [18].

3.4   FEATURE SELECTION

Identifying the right features and extracting the same is the major role in object tracking. This selects the features which are very important and makes the object unique. It is done manually. Feature selection and object representation are strongly correlated. Contour representation uses object features like object edges [13]. Some are listed below:

3.4.1   EDGES

Edges are defined as the region or outline between the object and the background. An object boundary can be easily located using human eyes. Changes in intensity are strong near the boundaries. Change in intensity can be recognized using the edge feature and then the tracking algorithm tracks the objects. Edge features are not sensitive to change in light intensities [10].

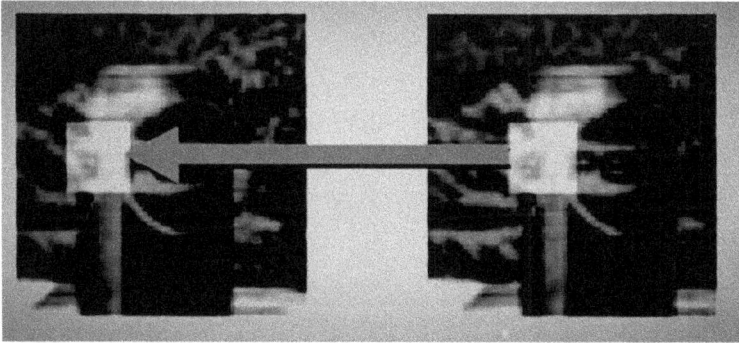

FIGURE 3.5 Location of the pixel changed but brightness remains same [20].

3.4.2 Optical Flow

Motion features are sometimes defined using brightness patterns in a visual scene. We experience this phenomenon in our daily life when driving and looking outside the window. The objects viewed from the car (streets, buildings, trees, etc.) looks like they are moving backward. Apparent motion is determined using pixels movement among frames. The limitation here is brightness, which leads to reliable brightness in different frames. Figure 3.5 illustrates this process.

3.4.3 Color

Diverse colors are used to store the different object feature information. Generally, color is described using RGB color space. Sometimes in computer vision it can be represented by YCbCr and HSV color space. The major problem of using color is its dependency to changes in illumination. Object color is not only affected by the illumination factor, it is also affected by the object reflection properties [13].

3.4.4 Texture

Intensity variation of a surface can be measured using regularity or smoothness properties. The object descriptors are generated using a pre-processing step [13]. Texture descriptors, for example, require the corresponding stages to level, edge, spot, etc. [14].

3.5 OBJECT TRACKING TECHNIQUES

Object tracking methods [21] are listed below:

    i. Point
   ii. Kernel
  iii. Silhouette

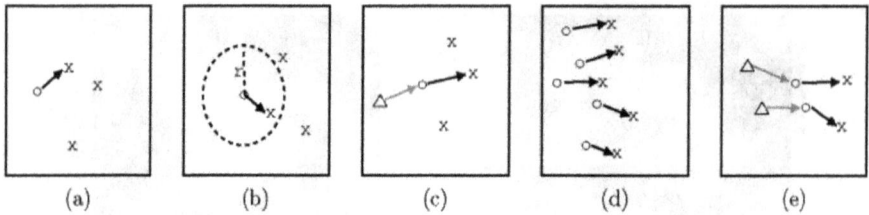

(a)	(b)	(c)	(d)	(e)

FIGURE 3.6   Point-based tracking [10].

3.5.1 POINT

The point method is an accurate method and the algorithm is reliable and robust [10]. The moving object can be identified with the features extracted. Here, the moving object is represented by its feature points (Figure 3.6). The principle used here considers objects in the consecutive frame and points on the previous frame are used [18]. It is able deal with very small objects. A disadvantage is the presence of false detections.

Some approaches of point tracking are as below:

  i. Kalman Filter

This filter is constructed based on the optimum recursive data processing algorithm [10]. It composes two steps: estimate and update. Next state can be predicted with current state and update the estimated measurements. It is able to do the following:

- Optimal solutions
- Noise handling
- Tracks the single and multiple objects

  ii. Particle Filter

The particle filter is able to track under linear or non-linear conditions [19]. It is able to produce better results than other methods [19]. The procedure is described below:

1. Generate samples to represent the initial probability.
2. Predict the next state using the prior equation.
3. Get the weights for the states computed using the observation. (from step 2). Predicted states along with the weights collectively represent the state distribution.
4. Resample the distribution to achieve the uniformly distributed current state omitting the least-significant representation.
5. Continue Steps 2 through 4 until all the observations are exhausted.

  iii. Multiple Hypothesis Tracking

Motion from several frames is combined in multiple hypothesis tracking (MHT). It produces good results if correspondence is established by observing several frames instead of considering the successive frames. MHT provides several features for an object. Tracking algorithm over a period is an iterative algorithm [22].

3.5.2 KERNEL-BASED TRACKING APPROACH

This approach is mainly based on motion (parametric) and the origin of the object. It is computed using the subsequent frames [19]. Another advantage of this method is that it easily predicts its estimation. A trajectory of an object is required to find whether the object is stationary or moving. Identifying the object region covering the object is very important [13].

Kernel refers to the appearance or shape of the object. Various original shapes are available (rectangular and ellipse), to represent the object [23]. Kernel tracking has four types. Every method differs on the representation and number of objects.

i. Simple Template Matching

This is a [19, 24] fundamental procedure for investigating the particular area in the video frame.

A reference frame is compared with the target frame in a video sequence. A single object can be tracked within a video and semi overlapping is done. This method helps to find even small components in an image that match with every successive frame.

- Single object tracking.
- Incomplete occlusion.
- Need of an external initialization.

ii. Mean Shift Method

The main objective of this method [13] is to identify a particular part in a video frame that is most identical to the model previously prepared. Histogram representation is used for tracking. The gradient ascent method can also be used for tracking and provides similar results. The target object can be tracked using rectangular or elliptical structures. The target model can be framed using a PDF function. An asymmetric kernel has been used to regularize the target model.

3.5.3 SILHOUETTE APPROACH

Objects with complex shapes can't be correctly characterized. Silhouette methodology will give accurate shape sketch of an objects. A tracking model can be developed by finding the region of the object based on past frames. It is classified into two classes: i. shape matching, and ii. contour tracking. It has the following advantages.

i. It supports many objects shapes
ii. It handles object occlusion
iii. It deals with splitting and merging

i. Contour Tracking

Contour tracking is also called boundary tracking. It can create a unique contour from the previous frame and current frame. Contour tracking is similar to state space models [22]. Edge-based features are used because they are unaffected to lighting conditions. This creates as strong contour. Since the object boundary is small, its speed increases. It has two techniques. The first one uses the state space model, which increases the shape and motion, and the other one directly grows the contour. A gradient descent procedure is used because that maximizes a similarity score between the model and the current image region.

ii. Shape Matching

The current frame object model can be developed. Template and shape matching both perform the same. Shape matching also discovers the matching silhouettes. It is similar to point matching. It is capable of:

1. Tracking single objects using edge-based features.
2. Using the Hough transform to solve the occlusion.

3.6 INTRODUCTION TO DEEP AND MACHINE LEARNING TECHNIQUES

Both machine learning (ML) and deep learning (DL) are subdivisions of artificial intelligence. The computer or machine performs specific actions based on the object patterns and inferences without clear instructions. For this an algorithm and a model has to be carefully developed. The model becomes efficient only when the training date set is huge [25, 26]. The model must be trained through iterations with the labeled training data set so as to produce the output. Once the training is over the model can be tested with unlabeled data. The term ML was created by Arthur Samuel [27]. Alan Turing proposed the question "Can machines think?"

Supervised learning of a mathematical model has been developed for a set of labeled data, whereas reinforcement methods deal with unlabeled training data. Examples of a classification algorithm are separating "spam" mails. Regression uses continuous outputs, e.g., frequency, voltage, or product prize. Unsupervised learning method works with training data that does not have a label. If the application is a development of a robot, the procedures will have its own procedures from previous learning. Dimensionality reduction is one of the techniques used to reduce the number of features.

3.6.1 DEEP LEARNING VS MACHINE LEARNING TECHNIQUES

Modern ML tools incorporate neural networks in the sequence of layers to learn from the training data set. Computational intelligence (CI) has been developed as a powerful method for making a machine learn. It works well in the field of neural networks, fuzzy systems, and evolutionary algorithms. Recent advances in DL have been playing an important role in dealing with huge amounts of unlabeled data. Due to the remarkable successes of DL techniques, we are now able to boost quality of service (QoS) significantly. More complex features can be easily extracted by deep neural network (DNN) (also called deep belief network and convolutional neural network (CNN)) and used to efficiently learn their representations. However, implementing deep learning faces many implementation challenges such as large data sets (needed to ensure desired results), high complexity of the network, high computational power, etc., which need to be addressed to effectively implement deep learning to solve real world image processing problems.

3.7   EXAMPLES

3.7.1   Vehicle Detection

Vehicle detection uses computer vision technique for tracking vehicles. Vehicle detection has an important role in autonomous driving applications like forward collision detection, adaptive travel control, and automatic lane keeping. The following example shows vehicle detection using the DL method. Deep learning is the dominant tool that automatically extracts image features. The vehicle data set is loaded and the convolution neural network is designed. First the inputs are split into training (60%) and testing (30%). This example uses 295 vehicle images. To get accurate classification more data are required. Each image has a label value. An example image from the training data [27] set is displayed in Figure 3.7.

A CNN forms the foundation for the R-CNN detector [28]. It can be created using the neural network toolbox. It has three layers: input, hidden, and output. However, for detection, input size must be small. Here the object size is greater than [13], so the input size is selected as. Next we have to decide the middle layers of CNN. It contains a number of layers such as convolutional, rectified linear units, and pooling. These are the fundamental blocks of CNN. The final output layers contain fully connected and a soft-max layer.

Figure 3.8 shows testing a single image and provides a good promising result. The detector performance can be measured by considering the entire dataset to evaluate the performance of the detector test with the entire dataset. Computer vision toolbox in MATLAB provides this feature [27]. Precision is shown in Figure 3.9.

3.7.2   Training of a Cascade Detector

The vision cascade object detector has many pre-trained classifiers for detecting frontal regions and the upper part of the body. The kind of object that can be detected

FIGURE 3.7   Training image from the data set.

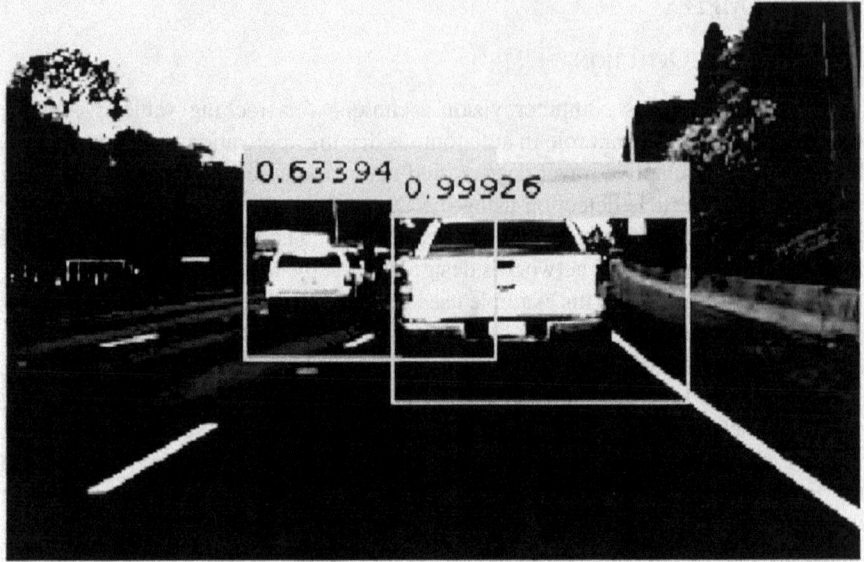

FIGURE 3.8 Testing with single image.

with this category includes objects with a constant aspect ratio. Aspect ratios of faces, stop signs, and cars are constant. The cascade detector contains a window that slides over an entire image. Presence of the object can be estimated using the cascade detector. Since the aspect ratio of 3D object is not fixed, the detector is particular in plane rotation. A single detector cannot detect an object in all three dimensions. Negative samples should be rejected as soon as possible.

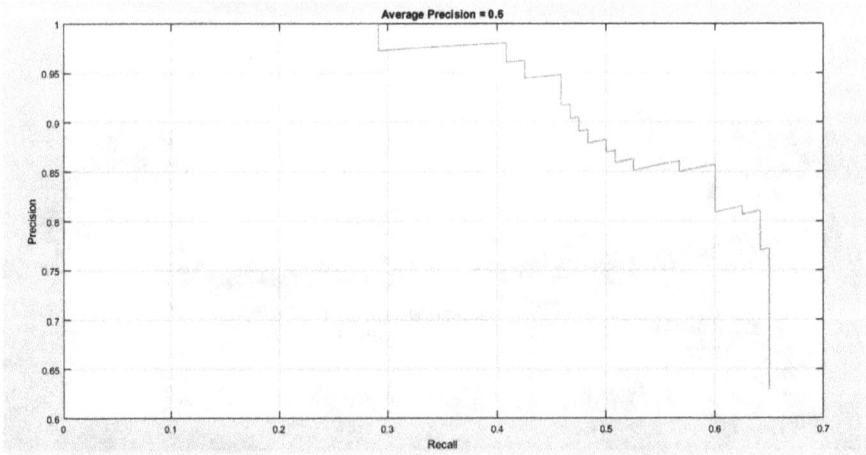

FIGURE 3.9 Average precision.

FIGURE 3.10 Visualization of the HOG features of a bicycle.

3.7.3 FEATURE TYPES AVAILABLE FOR TRAINING

According to ML and pattern recognition, a feature is an individual measurable property or characteristic of a phenomenon being observed. Histogram of oriented gradients (HOG) features is used to detect objects. HOG, is a feature descriptor that is often used to extract features from image data. The outline of a bicycle can be viewed using HOG features (Figure 3.10).

3.7.3.1 Supply Positive Samples

The Training Image Labeler app is used to create positive samples easily (Figures 3.11 and 3.12).

FIGURE 3.11 Positive samples.

FIGURE 3.12 Stop sign as positive example.

Positive samples can be supplied used the following methods:

1. Identify rectangular regions.
2. Crop the object of interest.

3.7.3.2 Supply Negative Images

The train cascade object detector does not specify negative samples.

Region of interest can be obtained by defining the type of detector input image and region of interest correctly [28–32] (Figures 3.13 and 3.14)

FIGURE 3.13 Face detection.

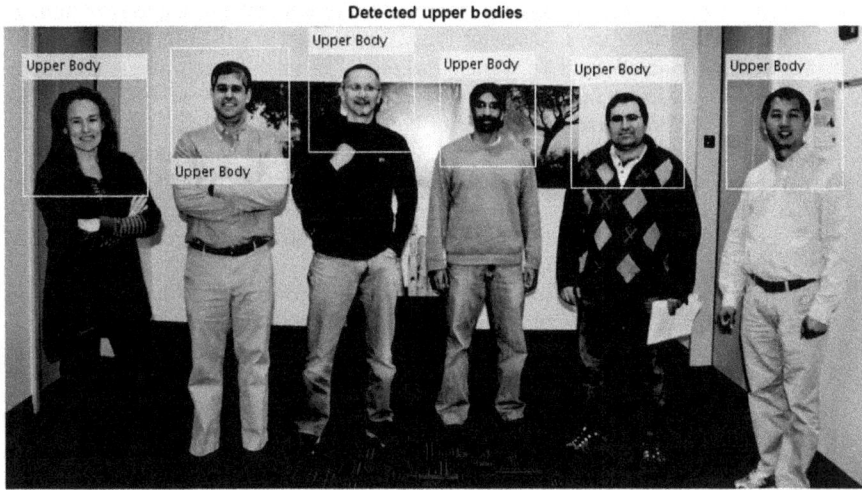

FIGURE 3.14 Upper portion of body detection in an image.

1. Construct the cascade object detector.
2. Call the cascade object detector for the input image

Region of interest can be obtained by defining the type of detector input image and region of interest correctly [28–32].

3.8 CONCLUSION

This chapter provides a thorough survey on visual-tracking method. A detailed study of the connecting methods is also discussed. Moreover, different methods adopted for visual tracking are introduced. The chapter highlights the features of algorithms for researchers in the field of visual tracking. Future work focuses on object detection with a non-static background and having multiple cameras which can be used in real-time surveillance applications.

REFERENCES

1. Dana H. Ballard and Christopher M. Brown. Computer Vision. Prentice Hall, 1982. ISBN 978-0-13-165316-0.
2. T. Huang and Carlo E. Vandoni, (ed.). Computer vision: Evolution and promise (PDF). 19th CERN School of Computing. Geneva, CERN, pp. 21–25, 1996. doi:10.5170/CERN-1996-008.21. ISBN 978-9290830955.
3. Milan Sonka, Vaclav Hlavac, and Roger Boyle. Image Processing, Analysis, and Machine Vision. Thomson, 2008. ISBN 978-0-495-08252-1.
4. Reinhard Klette. Concise Computer Vision. Springer, 2014. ISBN 978-1-4471-6320-6.
5. Linda G. Shapiro, and George C. Stockman. Computer Vision. Prentice Hall, 2001. ISBN 978-0-13-030796-5.
6. Tim Morris. Computer Vision and Image Processing. Palgrave Macmillan, 2004. ISBN 978-0-333-99451-1.

7. Bernd Jähne and Horst Haubecker. Computer Vision and Applications, A Guide for Students and Practitioners. Academic Press, 2000. ISBN 978-0-13-085198-7.
8. Jae-Yeong Lee and Wonpil Yu. Visual tracking by partition-based histogram backprojection and maximum support criteria. Robotics and Biomimetics (ROBIO), 2011 IEEE International Conference on, 7–11 December, pp. 2860, 2865, 2011.
9. Grandham Sindhuja and Renuka Devi. A survey on detection and tracking of objects in video sequence. International Journal of Engineering Research and General Science, 3(2), 2015. ISSN2091-2730
10. Alper Yilmaz, Omar Javed, and Mubarak Shah. Object tracking: A survey. ACM Computing Surveys (CSUR), 38(4):13, 2006.
11. Anshul Vishwakarma and Amit Khare, "Vehicle detection and tracking for traffic surveillance applications: A review paper", IJCSE, 6(7), 2008, ISSN 2347-2693.
12. Sanna Agren. Object tracking methods and their areas of application: A meta-analysis. A thorough review and summary of commonly used object tracking methods, 2017.
13. Sandeep Kumar Patel and Agya Mishra. Moving object tracking techniques: A critical review. Indian Journal of Computer Science and Engineering, 4(2):95–102, 2013.
14. Pennsylvania State University. Probability density functions. https://onlinecourses.science.psu.edu/stat414/node/97, 2016. [Online; accessed 06 December 2016].
15. Michael J Black and Allan D Jepson. Eigentracking: Robust matching and tracking of articulated objects using a view-based representation. International Journal of Computer Vision, 26(1):63–84, 1998.
16. Rupali S. Rakibe and Bharati D. Patil. Background subtraction algorithm based human motion detection. International Journal of Scientific and Research Publications, 3(5), May 2013, ISSN 2250–3153.
17. K. Srinivasan, K. Porkumaran, and G. Sainarayanan. Improved background subtraction techniques for security in video applications. International Conference on Anticounterfeiting, Security, and Identification in Communication, 2009, ISSN: 2163-5048.
18. Sen-Ching S. Cheung and Chandrika Kamath. Robust techniques for background subtraction in urban traffic video.
19. Joshan Athanesious J and Suresh P. Implementation and comparison of kernel and silhouette based object tracking. International Journal of Advanced Research in Computer Engineering & Technology: 1298–1303, March 2013.
20. Min Sun and Krstic Srdjan. Optical flow. http://www.cs.princeton.edu/ courses/archive/fall08/cos429/optiflow.pdf, 2016. [Online; accessed 06 December 2016].
21. Nirav D. Modi. Moving object detection and tracking in video. International Journal of Electrical, Electronics and Data Communication, 2(3), March 2014, ISSN 2320–2084.
22. J. Joshan Athanesious and P. Suresh. Systematic survey on object tracking methods in video. International Journal of Advanced Research in Computer Engineering & Technology (IJARCET), 242–247, October 2012.
23. R. Hemangi Patil and K. S. Bhagat. Detection and tracking of moving object: A survey. International Journal of Engineering Research and Applications, 5(11):138–142, 2015.
24. S. Saravanakumar, A. Vadivel, and C.G. Saneem Ahmed. Multiple human object tracking using background subtraction and shadow removal techniques. Signal and Image Processing (ICSIP), 2010 International Conference on 15–17 December, vol., pp.79, 84, 2010.
25. R. John, Forrest H. Bennett, David Andre, and Martin A. Keane. Automated design of both the topology and sizing of analog electrical circuits using genetic programming. Artificial Intelligence in Design '96. Springer, Dordrecht. pp. 151–170, 1996.
26. C. M. Bishop. Pattern Recognition and Machine Learning, Springer, 2006. ISBN 978-0-387-31073-2.
27. Mathworks.com.

28. Shaoqing Ren, et al. Faster R-CNN: Towards real-time object detection with region proposal networks. Advances in Neural Information Processing Systems, 2015.
29. R. Lienhart, A. Kuranov, and V. Pisarevsky. Empirical analysis of detection cascades of boosted classifiers for rapid object detection. Proceedings of the 25th DAGM Symposium on Pattern Recognition, Magdeburg, Germany, 2003.
30. Ojala Timo, Pietikäinen Matti, and Mäenpää Topi. Multi-resolution gray-scale and rotation invariant texture classification with local binary patterns. In IEEE Transactions on Pattern Analysis and Machine Intelligence, 24(7):971–987, 2002.
31. H. Kruppa, M. Castrillon-Santana, and B. Schiele. Fast and robust face finding via local context. Proceedings of the Joint IEEE International Workshop on Visual Surveillance and Performance Evaluation of Tracking and Surveillance, pp. 157–164, 2003.
32. Marco Castrillón, Oscar Déniz, Cayetano Guerra, and Mario Hernández. ENCARA2: Real-time detection of multiple faces at different resolutions in video streams. Journal of Visual Communication and Image Representation, 18(2):130–140, 2007.

4 Fuzzy MCDM: *Application in Disease Risk and Prediction*

Rachna Jain, Abhishek Kathuria, and Devanshi Mukhopadhyay
Bharati Vidyapeeth's College of Engineering

Meenu Gupta
Chandigarh University

CONTENTS

4.1 FUZZY MCDM: INTRODUCTION

Fuzzy multi-criteria decision-making (MCDM) as a tool has been quite useful where data is huge, incomplete, and uncertain. Fuzzy-based analysis approach has been used for data mining techniques to identify patterns and formulate equations and decision trees when the data is huge, contains uncertainties, and where traditional forms of statistical analysis have failed. Medicine is one such field where a huge amount of raw, uncertain data is available. Different diseases manifest differently in patients, that is, symptoms are similar, but the quantitative and qualitative aspect of the severity of symptoms may differ, leading to uncertainty [1]. MCDM was proposed in the 1970s and it is a subdiscipline of operations research that explicitly evaluates multiple conflicting criteria in decision-making [2].

Incorporating fuzzy logic with MCDM techniques significantly enhances the quality and adaptability of decision-making, generating probabilistic values even when data is ambiguous and uncertain. Fuzzy MCDM is used as an alternative decision-making tool on multiple-criteria problems by decision-makers, where the linguistic values are represented as a fuzzy numbers and are responsible for measuring or evaluating the importance of the criteria. Although MCDM is a novel approach for solving problems with multiple conflicting criteria, it is still not implemented for decision-making on a managerial level due to certain drawbacks [3]. One of these drawbacks is that MCDM does not allow interdependency of criteria, which makes the solution obtained by the algorithms unfeasible. Unlike a strict hierarchy system, autonomous decision-makers are inclined to use more than the required number of criteria [4]. The purpose of fuzzy MCDM is to design mathematical computational tools for quantitative and qualitative evaluation of multiple and inconsistent criteria. The evaluation criteria help the decision-makers in ranking, classifying the alternatives, and choosing the best path forward. Many methods and tools are used for disease prediction and risk analysis using fuzzy MCDM. Applying and developing a fuzzy MCDM expert system for disease risk prediction has many merits, a primary one being that of preemptive mitigation steps to lessen the financial burden upon the patient [5]. The different expert system of risk prediction for diseases like heart diseases, breast cancer [6–9], and diabetic retinopathy have been developed and tested by various individuals. These systems do prove to be a viable tool for disease prediction. In this work, the architectural composition of a general fuzzy system is explored to understand how the patient's biological data can be applied to a system as fuzzy inputs to get a crisp value output of the likely disease he/she could suffer. A new method that combines fuzzy logic with the multi-criteria decision-making system is meticulously explored. Three different case studies of applying fuzzy MCDM to disease prediction are explored and evaluated based on other state-of-the-art methods, i.e., neural networks.

This chapter is further divided into sections are proposed as: Section 4.2 describes the literature review done by many researchers, Section 4.3 focuses on the approach/methodology used in this chapter, and Section 4.4 discusses the case studies on MCDM. The chapter concludes with Section 4.5 with its future scope.

4.2 LITERATURE REVIEW

Energy has a significant role in the growth of industries, technologies, and agriculture. In other ways, a country's development (i.e., social and economic) is highly dependent on energy planning, which is a big issue for every countries. These problems are evaluated using MCDM methods. Many researchers study using the MCDM method to resolve the above mentioned problem.

In [10], I. Kaya et al. present a cumulative study on fuzzy MCDM: application and methodologies discussed by different researchers in the field of energy. They studied many research papers (who works on fuzzy MCDM method) written by different researchers to solve energy decision-making problems including parameters such as year, types of fuzzy sets, country, fuzzy MCDM method, journal, and document type. They conclude that Turkey and China are the countries where good numbers of research in field of energy-related problems (using fuzzy MCDM methods) were published. In [11], M. Cloak et al. proposed a fuzzy set based integrated MCDM model for prioritization of renewable energy alternatives (in Turkey). Authors used two methods (type-2 fuzzy AHP (analytic hierarchy process) and hesitant fuzzy TOPSIS) to analyze the weight of decision tree and prioritize the renewable energy alternatives. Further in [12], M. Yucesan et al. proposed a multi-phase MCDM model for selection of green suppliers. Weight selection measures for green suppliers are analyzed using best-worst method (BWM) and order ranking is analyzed by interval type-2 fuzzy technique (IT2F TOPSIS). This proposed model is implemented as a selection process for green in a plastic injection molding facility (in Turkey). In [13], C. N. Wang et al. consider a problem that occurs in the garment industry of Vietnam. In this work, the authors implemented a MCDM model to enhance the selection and evaluation process for suppliers of garment industries. Some factors considered in this study were the selection of supplier as determined by a triple bottom line (TBL) model, weight of all factors as determined by the fuzzy analytical hierarchy process (FAHP) method, and TOPSIS to rank the best supplier of fabric in garment industry. The authors concluded that Decision Making Unit 10 was the best method for finding the ranking of a supplier. In [14], Y. Suh et al. worked with a new MCDM method using the Kriterijumska Optimizacija I Kompromisno Resenje (VIKOR) approach. This proposed approach evaluated the service quality of mobile for subjective (used DEMATEL method) and objective weights (used Shannon Entropy). The authors concluded that the proposed model was capable of analyzing mobile services on the basis of overall performance.

In [15], Kirti Sharawat et al. used MCDM method for determining the highest quality diet suitable for the diabetic patient. The categorization of the diet is done based on categories such as calories, body fat, and carbs. This work used an analytic hierarchical process model along with the MCDM approach for ranking the diet suitable for diabetic patients. Fuzzy TOPSIS method was used for the validation part which was developed by Wang et al. [16]. They concluded that, F1 (a food type) showed better results than others. Li-En Wang et al. [17] used a new hybrid model with a combination of IVIFSs and COPRAS methods. They applied fuzzy MCDM method on a hybrid model to find and rank the risk of failure modes. The weight of risk factors was calculated by a method named IVIF-ANP by taking into consideration individual relationships. This model turned out to be more accurate and more efficient than the traditional FMEA model. A fuzzy MCDM approach was discussed by Chih-Young Hung et al. for evaluating four account receivable (A/R) collection of instruments [18]. For each case, the determination of decision criteria weights were calculated with the help of fuzzy analytic (a hierarchy process) for dealing with the subjective judgment's qualitative attribute. The fuzzy MCDM approach enabled for efficient decision-making and synthesizing the group decision. Lazim Abdullah [19] provides a review of the various classification techniques of fuzzy MCDM. Abdullah also discusses the real-life applications of fuzzy MCDM in Malaysia.

Andrzej Piegat et al. proposed an approach for analyzing the severity of chronic liver disease using fuzzy MCDM techniques. The proposed characteristic objects method (COMET) approach was compared to the TOPSIS and AHP methods [20]. The results obtained by Piegat showed that the COMET approach gave better results as compared to the other two approaches, which was evident from the fact that the COMET approach gives 966,309 correct answers as compared to 805,345 correct answers given by TOPSIS and 933,165 correct answers given by AHP. The main implementation suggested by Cheng was an evaluation of a weapons system [21]. The two parameters for evaluating desired weapons systems in the real-world conflict in nature, that is, a tradeoff between the performance of the parameters and weapons system descriptions that are generally vague and linguistic in nature. The first problem can be solved via general MCDM techniques like AHP. For solving the second problem, Fuzzy MCDM techniques are needed. Cheng devised a general algorithm in which a judgment matrix was built by pairwise comparison of triangular fuzzy numbers, varying between 1 and 9. The eigenvectors that are present in the judgment matrix are estimated. The next section discusses the methodologies used in this chapter in detail.

4.3   METHODOLOGIES

This section discusses the mathematical procedure of fuzzy MCDM and architecture of fuzzy MCDM inference systems.

4.3.1   MATHEMATICAL PROCEDURE FOR APPLYING FUZZY MCDM

4.3.1.1   Fuzzification

Fuzzification is the term used to convert crisp input values into fuzzy input values [22]. The fuzzification process is responsible for converting scalar value into fuzzy value. It converts crisp values to membership values corresponding to the fuzzy set of the linguistic term [23]. Membership's functions are of various types like Gaussian waveform, triangular waveform, trapezoidal, etc. A Gaussian waveform is suitable for a system that demands high accuracy, while triangular and trapezoidal functions are more suited for systems where in a short period of time significant dynamic variations occur.

4.3.1.2   Fuzzy Sets

Fuzzy means nothing or that cannot be predicted, for example, weather. 'Fuzzy set' was introduced by Lotfi A. Zadeh and Dieter Klaua in 1965. In fuzzy sets, every element has different degrees of membership ($\mu_A(x)$) on a particular set A and element x whose limit range is between 0 and 1. In this, elements work on partial membership on a particular set [24], for example range can be between 0.0 and 1.0. When the element's value belongs to 0.0 it shows absolutely false and 1.0 means absolutely true. Fuzzy sets are represented with a tilde (~) symbol.

4.3.1.3   Crisp Sets

A Crisp set is also known as a classical set. It is a collection of distinct objects in which elements have complete membership (X or U) on that particular set, for example, students passed grades or not. The range of membership is belongs to either 0 or 1. There is no partial membership available for the element set.

4.3.1.4   Membership Functions

A membership function (MF) is outlined as a degree of truth [25]. The representation of member function is shown in eq. (1). It's a curve-like structure in which each input and output is mapped between values zero and one. Degree of membership is a value which is associated with the corresponding input value (i.e., the output of the membership function). Membership operator/value is shown in Figure 4.1.

$$\mu_A : X \rightarrow [0, 1] \tag{1}$$

4.3.2   GENERAL ARCHITECTURE OF A FUZZY MCDM INFERENCE SYSTEM

A fuzzy inference system (FIS) is used to map the fuzzy input set to the fuzzy output [24], shown in Figure 4.2. It is a way of applying human language reasoning with the concept of fuzzy logic incorporated in it.

For example,

If A, then B.
Therefore, B.

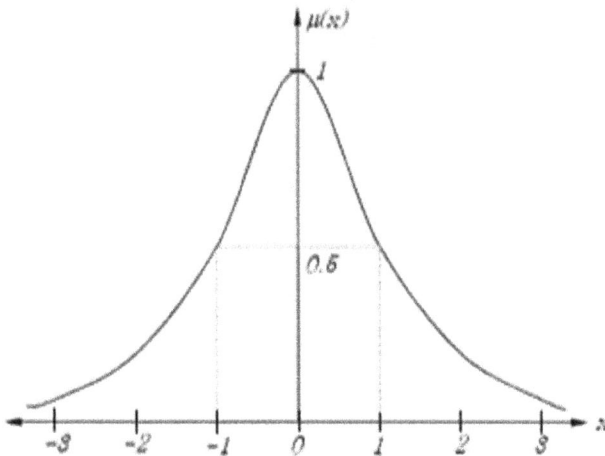

FIGURE 4.1   Membership function of fuzzy set [26].

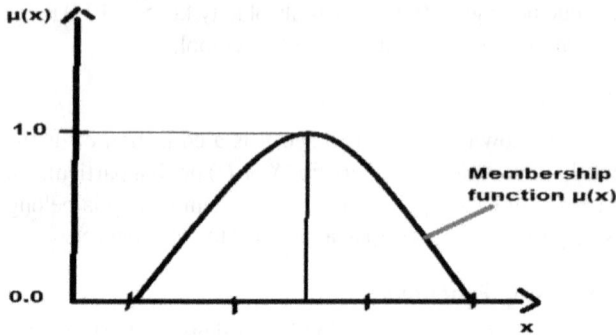

FIGURE 4.2   Gaussian membership function.

This form of reasoning is fairly strict, that is, B can only be if A. Fuzzy logic loosens this strictness by implying that B can be mostly if A is mostly that is to say

    If *A* then *B*
    Mostly *A*
    Therefore mostly *A*

4.3.2.1   Fuzzy If-Then Rules

In this case, when one condition is true then other condition will show the results. For example: It is of the following form: If 'a' is M, then 'b' is N. The antecedent is 'a' is M; the consequence is 'b' is N.

M and N are defined by fuzzy sets on the universe of discourses R and S. Here, 'a' and 'b' refers to the input and output respectively [27]. The implication of 'is' work different in both the antecedent and consequence, in the antecedent 'is' is used to assign a value between 0 and 1 while in the consequence 'is' is used to assign N to 'b'.

FIS architecture is shown in Figure 4.3. FIS includes four modules: fuzzifier, defuzzifier, inference engine (IE), and fuzzy knowledge base (FKB).

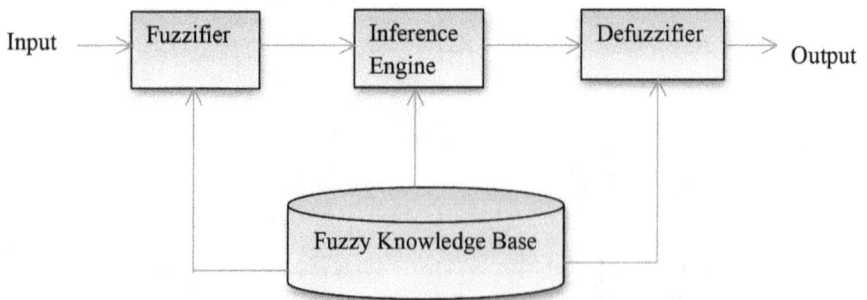

FIGURE 4.3   Fuzzy inference system architecture.

The fuzzifier converts the input crisp values of the antecedent to fuzzy sets. The inference engine gives the fuzzy output from a fuzzy input and the fuzzy knowledge base is a database that consist the rule base. The defuzzifier converts the fuzzy set obtained by the inference engine to a crisp value [28]. The next section discusses case studies for diagnosing heart disease (a medical problem) using the fuzzy MCDM method.

4.4 CASE STUDIES

A huge amount of financial resources is spent on identification disease and diagnosing it, but the main problem come in identification of the disease, which is time-consuming and complex, leading to further health problems in the patient [29]. Using medical records containing numerous factors like cholesterol level, blood pressure, etc., an expert system can be developed to evaluate the risk of certain diseases of the patient, thus enabling doctors and the patient to take certain preemptive measures [30]. This would not only lessen the financial burden on the patient but also promote a healthier lifestyle for them. Further, the professional burden on the doctors would be reduced, which would enable them to devote more time to medical research for further development of the knowledge base of the expert system.

This section discusses some proposed fuzzy MCDM systems developed by various individuals for disease risk prediction and analysis.

4.4.1 DETECTION AND RISK PREDICTION OF MEDICAL PROBLEM (HEART DISEASE)

The dataset used was UCI cardiovascular disease diagnosis, Cleveland Heart disease dataset [31]. The collected dataset was huge (i.e., 76 attributes) with a great deal of irregular/unmanaged data. The data was first pre-processed and cleaned. After that, the dataset was converted into CSV format. The redundant features in the dataset were removed using data reduction techniques to get a total of six attributes as depicted in Figure 4.4. These attributes are mentioned in the input field column of Table 4.1. Fuzzification of the input variable was applied using membership functions like triangular and trapezoidal.

Figure 4.4 shows the medical diagnosis processing flowchart, where the heart diseases dataset (UCI Cleveland) is considered. Initially, the dataset was not in structured form, the pre-processing was applied to convert the unstructured dataset into structured form or processing form. After pre-processing, attributes are selected with the help of a data mining technique. Next, fuzzification processes are applied to input data and generate some rules with the help of fuzzy inference rule system using MATLAB. The defuzzification process is used to get output based on the antecedent, i.e., fuzzy input. The rules were applied in MATLAB R2012a in the rule editor. The Mamdani inference system was used, and finally the output was defuzzified to obtain the crisp output. This fuzzy MCDM based system was examined with Neural Network and J48 Decision Tree, where training and testing data was split in the ratio of 60:40.

TABLE 4.1

Risk Factors and Ranges

Input Field	Range	Fuzzy Sets
Age	Less than 38	Young
	Between 33 and 45	Middle
	Greater than 45	Old
Blood Pressure	Less than 138	Low
	Between 126 and 152	Medium
	Between 142 and 172	High
	Greater than 154	Very High
Cholesterol	Less than 197	Low
	Between 188 and 250	Risk
	Greater than 217	Terrible
Heart Rate	Less than 141	Low
	Between 111 and 194	Medium
	Greater than 152	High
Old Peak	Less than 2.5	Low
	Between 1.5 and 4.2	Risk
	Greater than 2.5	Terrible
Thallium Scan	3	Normal
	6	Fixed Defect
	7	Reversible Defect

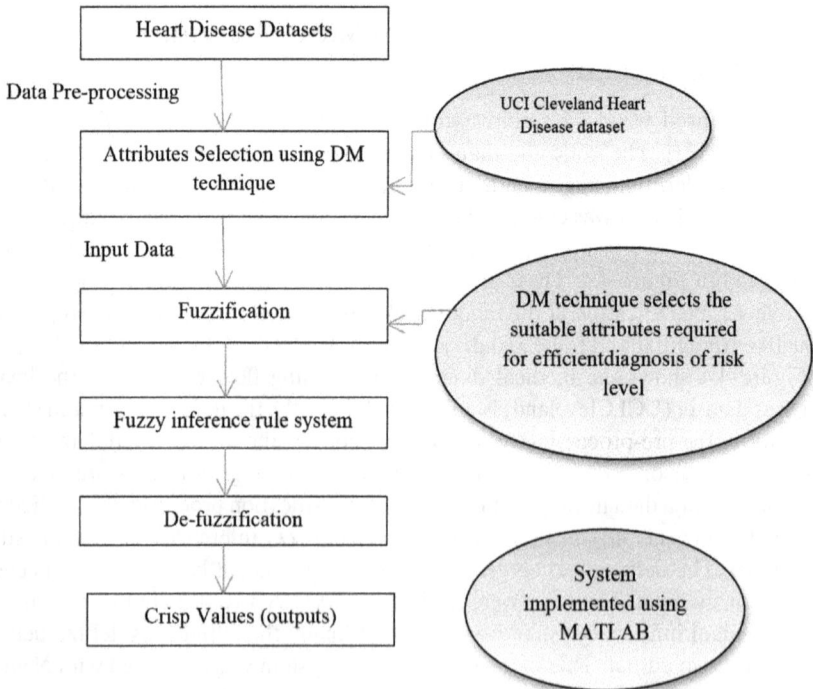

FIGURE 4.4 Flowchart diagram for processing medical diagnosis.

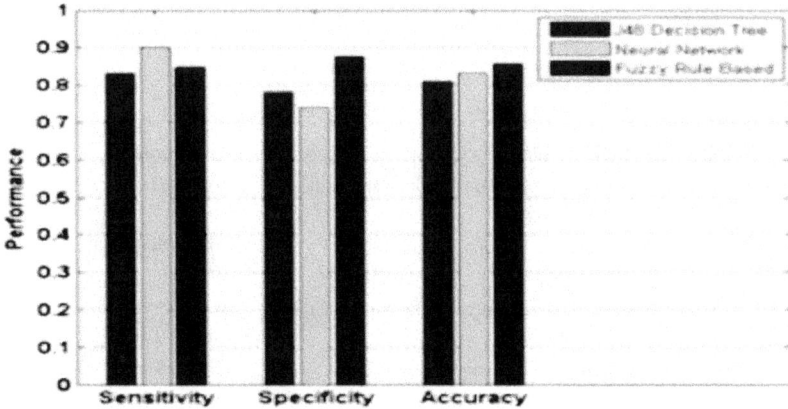

FIGURE 4.5 Training dataset performance.

By observing Figures 4.5 and 4.6, it was seen that the performance fuzzy MCDM system developed to predict heart disease is quite close to that of the Neural Network with close to 80% accuracy.

4.4.2 RISK PREDICTION OF BREAST CANCER USING FUZZY LOGIC

Records from 2001 to 2006 from four institutes, the National Cancer Research Programme (NCRP), Government Hospitals of Coimbatore, VNCC (Cancer Centre), and the Department of Oncology, Surgical Oncology and Radiation Oncology, Sri Ramakrishna Hospital, were used as the datasets for the case study [32]. After removing erroneous and incomplete data, a total of 12,595 records of cancer were

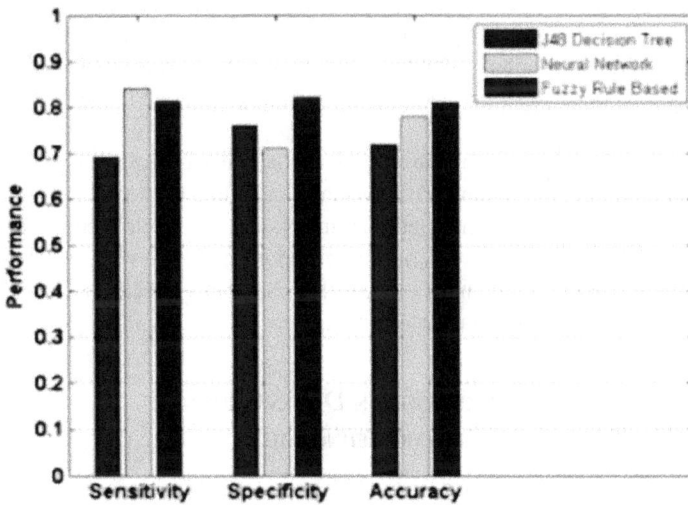

FIGURE 4.6 Testing dataset performance.

TABLE 4.2
Inference Rules

Serial Number	If-Then Rule	Implication
1	If Node = 1 and Tumor = very small, then Metastasis = no	Serious (Stage II A)
2	If Node = 1 Tumor = small, then Metastasis = no	Serious (Stage II B)
3	If Node = 2 and Tumor = medium, then Metastasis = no	Serious Moderate (Stage III A)
4	If Tumor = large and Node = 2, then Metastasis = no	Serious Moderate (Stage III B)
5	If Tumor = medium and Node = 3, then Metastasis = no	Very Serious Moderate (Stage III C)
6	If Tumor = large and Node = no, then Metastasis = no	Very Serious Moderate (Stage III B)
7	If Tumor = large and Node = 3, then Metastasis = no	Very Serious Moderate (Stage III C)
8	If Tumor = small and Node = 1, then Metastasis = yes	Very Serious (Stage IV)

recorded; out of these records 1,862 records accounted for breast cancer. Out of the 1,862 records of breast cancer, 181 records were found with complete historical details of the breast cancer patients. The parameters used for the analysis are breast cancer tumor size, number of Lymph nodes, and metastasis.

A fuzzy MCDM system was incorporated using ID3 algorithm and integrated with MATLAB environment. The linguistic variables very serious moderate (VSM), very serious (VS), not serious (NS), and serious (S) were used for fuzzification of the above parameters. Mamdani inference system was used as inference mechanism. Eight inference rules were defined and used. These generated rules are shown in Table 4.2.

The results obtained are compared with the actual medical data as depicted in Table 4.3, and the performance of this system is calculated.

From Table 4.4 it can be concluded that this system is a viable method for observing biological data and risk prediction based upon this data, whether this data be vague and ambiguous. In the next section, this chapter discusses the application of fuzzy logic for determining diabetic retinopathy.

4.4.3 APPLICATION OF FUZZY LOGIC IN DETERMINATION OF HARD EXUDATES IN DIABETIC RETINOPATHY

Diabetic retinopathy is a disease that arises in the eyes of people having type 1 or type 2 diabetes [33]. This case study shows a way to automatically detect the existence of standard diabetic retinopathy (DIARETDB0) dataset in the colored fundus

TABLE 4.3

Comparison of Obtained Results with Actual Data Used

Tumor	Lymph Node	Metastasis	T	N	M	Risk	Value	Risk
1	0	0	<2 cm	-	-	NS	4.4	NS
2	0	0	>2–5 cm	-	-	S	15	S
1	1	0	<2 cm	1–3 nodes	-	S	20	S
1	2	0	>2 cm	4–9 nodes	-	VSM	25	VSM
3	0	0	>5 cm	-	-	S	15	S
0	2	0	-	4-9 nodes	-	VSM	25	VSM
2	1	0	>2–5 cm	1–3 nodes	-	S	14.8	S
2	2	0	>2–5 cm	4–9 nodes	-	VSM	25	VSM
3	1,2	0	>5 cm	1–9 nodes	-	VS	34.8	VS
4	0,1,2	0	CW/S	>1 node	-	VS	34.8	VS
1–4	3	0	Any	>9 node	-	VS	34.8	VS
1–4	1–3	1	Any	>1 node	YES	VS	34.8	VS
0	1	0	-	1–3 nodes	-	S	18	S

images. This approach includes the application of fuzzy MCDM logic in the recognition of various characteristics of diabetic retinopathy in colored fundus pictures. Machine learning algorithms can precisely identify the retinal vasculature from these colored fundus images. Artificial neural networks are beneficial for automatic detection of diabetic retinopathy. Applying fuzzy logic on this set, various fuzzy rules are obtained and these rules are utilized in identifying the existence of diabetic retinopathy in these colored fundus images.

Color spaces are a way of relating color between machine and machine or person and person. The various color spaces used are: XYZ, YIQ, LUV, HSV, and Lab. There are no boundaries on a fuzzy set. They can variate from 'member of the set' to 'not a member of the set,' which is defined by the membership functions. Pliability can be given to fuzzy set models by using assertions such as 'light' or 'flat'. The fuzzy set F is given in equation 2:

$$F = \left\{ \left(x, mf(x) \right) \mid x \in S \right\} \tag{2}$$

TABLE 4.4

Relation of Range with Linguistic Variables

Range	:	Risk
0–10	:	NS (Not Serious)
11–20	:	S (Serious)
21–30	:	VSM (Very Serious Moderate)
31–40	:	VS (Very Serious)

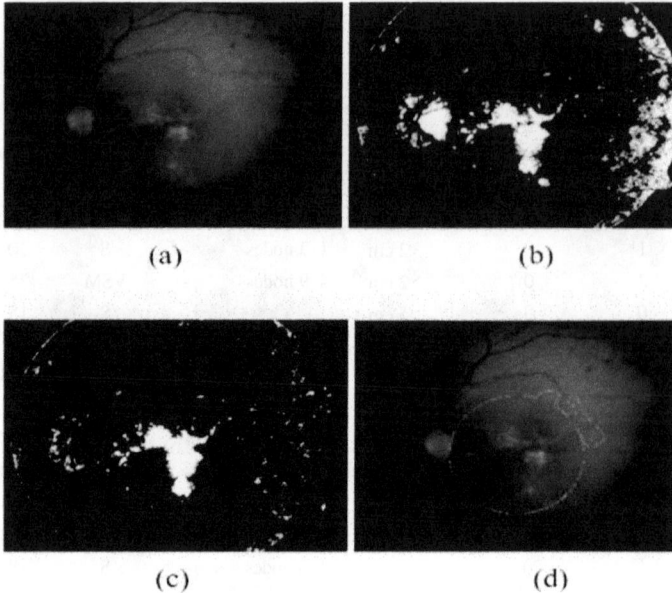

FIGURE 4.7 Identifying hard exudates in diabetic retinopathy: (a) actual image, (b) pre-processing, (c) segmentation, (d) imposition.

Where S is the set of objects and $mf(x)$ is known as membership function of x in object set S.

Now the fuzzy rules will be derived from these fuzzy sets. The syntax of fuzzy rules obtained is as follows:

IF Premise THEN Conclusion

Finally, the output obtained from these color space values is shown in Figure 4.7.

The result shows the average of the outputs of five color spaces. If the average is 1, then the patient is diagnosed as 'positive' and if the average comes out to be 0, the patient is diagnosed as 'negative'. Through this case study, it is concluded that diabetic retinopathy cannot be prevented. Regular eye exams, keeping a check on blood pressure, and early detection of the disease can prevent severe vision impairment. This will ultimately help medical practitioners identify this disease in its initial stages. The next section discusses the conclusion and future scope of this chapter.

4.5 CONCLUSION AND FUTURE SCOPE

In the above case studies, the practical application of a fuzzy MCDM system was observed and its value as a viable system for decision-making was demonstrated. The results of the proposed fuzzy MCDM system were comparable to that of other decision-making systems or in line with the biological/medical findings and conclusions. However, another field of study has developed by combining the features of neural networks and fuzzy logic. The fuzzy system that uses a neural network to determine its parameters after data processing is termed as neuro-fuzzy.

Like a neural network, a neuro-fuzzy inference system is divided into three or more layers. The first layer comprises of input processing. The second layer comprises determination and application of fuzzy rules in the form of weights assigned to each neuron and the third layer gives the output as a crisp value. As the neuro-fuzzy system is based on the fuzzy system, modifying the parameters of this system can prove to be difficult. It performs the function approximation by taking in training data. Scientists have discussed how neuro-fuzzy systems can be applied to the interpretation as well as diagnosis of diseases. Simulators for surgery can be developed using neuro-fuzzy systems. Simulators had a problem that they cannot recreate the feel of the elasticity of human tissue. This can be negated by implementing neuro-fuzzy systems in the simulators. Although neuro-fuzzy systems have their merits, there are certain problems still plaguing the actual implementation; e.g., many of the neuro-fuzzy systems would require more sophisticated and expensive hardware for more precise results. Hence, the economic viability of these systems come into play although they are advantageous the healthcare industry in general. Further, problems like the evaluation of system autonomy, reliability, and the precision of computation are also discussed.

REFERENCES

1. Jose, J. (2012, Oct). *ABS using fuzzy logic*. Accessed at: https://www.scribd.com/doc/109725294/ABS-Using-Fuzzy-Logic-Ppt.
2. Carlsson, C., & Fullér, R. (1996). Fuzzy multiple criteria decision making: Recent developments. *Fuzzy Sets and Systems, 78*(2), 139–153.
3. Mardani, A., Jusoh, A., MD Nor, K., Khalifah, Z., Zakwan, N., & Valipour, A. (2015). Multiple criteria decision-making techniques and their applications – A review of the literature from 2000 to 2014. *Economic Research-EkonomskaIstraživanja, 28*(1), 516–571.
4. Carlsson, C., & Fullér, R. (1996). Fuzzy multiple criteria decision making: Recent developments. *Fuzzy Sets and Systems, 78*(2), 139–153.
5. Adunlin, G., Diaby, V., & Xiao, H. (2015). Application of multicriteria decision analysis in health care: A systematic review and bibliometric analysis. *Health Expectations, 18*(6), 1894–1905.
6. Karabatak, M., & Ince, M. C. (2009). An expert system for detection of breast cancer based on association rules and neural network. *Expert Systems with Applications, 36*(2), 3465–3469.
7. Almurshidi, S. H., & Abu-Naser, S. S. (2018). *Expert System for Diagnosing Breast Cancer*. Al-Azhar University, Gaza, Palestine.
8. Abdel-Zaher, A. M., & Eldeib, A. M. (2016). Breast cancer classification using deep belief networks. *Expert Systems with Applications, 46*, 139–144.
9. Nilashi, M., Ibrahim, O., Ahmadi, H., & Shahmoradi, L. (2017). A knowledge-based system for breast cancer classification using fuzzy logic method. *Telematics and Informatics, 34*(4), 133–144.
10. Kaya, İ., Colak, M., & Terzi, F. (2019). A comprehensive review of fuzzy multi criteria decision making methodologies for energy policy making. *Energy Strategy Reviews, 24*, 207–228.

11. Çolak, M., & Kaya, İ. (2017). Prioritization of renewable energy alternatives by using an integrated fuzzy MCDM model: A real case application for Turkey. *Renewable and Sustainable Energy Reviews, 80*, 840–853.

12. Yucesan, M., Mete, S., Serin, F., Celik, E., & Gul, M. (2019). An integrated best-worst and interval type-2 fuzzy topsis methodology for green supplier selection. *Mathematics, 7*(2), 182.

13. Wang, C. N., Yang, C. Y., & Cheng, H. C. (2019). A fuzzy multicriteria decision-making (MCDM) model for sustainable supplier evaluation and selection based on triple bottom line approaches in the garment industry. *Processes, 7*(7), 400.

14. Suh, Y., Park, Y., & Kang, D. (2019). Evaluating mobile services using integrated weighting approach and fuzzy VIKOR. *PloS One, 14*(6), e0217786.

15. Sharawat, K., & Dubey, S. K. (2018). Diet recommendation for diabetic patients using MCDM approach. In *Intelligent Communication, Control and Devices* (pp. 239–246). Springer, Singapore.

16. Wang, T. C., & Lee, H. D. (2009). Developing a fuzzy TOPSIS approach based on subjective weights and objective weights. *Expert Systems with Applications, 36*(5), 8980–8985.

17. Wang, L. E., Liu, H. C., & Quan, M. Y. (2016). Evaluating the risk of failure modes with a hybrid MCDM model under interval-valued intuitionistic fuzzy environments. *Computers & Industrial Engineering, 102*, 175–185.

18. Hung, C. Y., Li, Y., & Chiang, Y. H. (2006, June). Evaluation of account receivable collection alternatives with fuzzy MCDM methodology. In *2006 IEEE International Conference on Service Operations and Logistics, and Informatics* (pp. 1009–1013). IEEE.

19. Abdullah, L. (2013). Fuzzy multi criteria decision making and its applications: A brief review of category. *Procedia-Social and Behavioral Sciences, 97*, 131–136.

20. Piegat, A., & Sałabun, W. (2015, June). Comparative analysis of MCDM methods for assessing the severity of chronic liver disease. In *International Conference on Artificial Intelligence and Soft Computing* (pp. 228–238). Springer, Cham.

21. Cheng, C. H. (1999). Evaluating weapon systems using ranking fuzzy numbers. *Fuzzy Sets and Systems, 107*(1), 25–35.

22. Tiwari, U. S. *Fuzzy sets (Type-1 and Type-2) and their applications.* Accessed at: https://www.iitk.ac.in/eeold/archive/courses/2013/intel-info/d1pdf3.pdf.

23. Yousefi, M., & Carranza, E. J. M. (2015). Fuzzification of continuous-value spatial evidence for mineral prospectivity mapping. *Computers & Geosciences, 74*, 97–109.

24. Bustince, H., Barrenechea, E., Pagola, M., Fernandez, J., Xu, Z., Bedregal, B., & De Baets, B. (2015). A historical account of types of fuzzy sets and their relationships. *IEEE Transactions on Fuzzy Systems, 24*(1), 179–194.

25. Alonso, S. K. *Mamdani's fuzzy inference method: Membership function.* Accessed at: http://www.dma.fi.upm.es/recursos/aplicaciones/logica_borrosa/web/fuzzy_inferencia/funpert_en.htm.

26. https://www.iitk.ac.in/eeold/archive/courses/2013/intel-info/d1pdf3.pdf.

27. Dvořák, A., Štěpnička, M., & Štěpničková, L. (2015). On redundancies in systems of fuzzy/linguistic IF–THEN rules under perception-based logical deduction inference. *Fuzzy Sets and Systems, 277*, 22–43.

28. Horgby, P. J., Lohse, R., & Sittaro, N. A. (1997). Fuzzy underwriting: An application of fuzzy logic to medical underwriting. *Journal of Actuarial Practice, 5*, 79–104.

29. Kumar, S., & Kaur, G. (2013). Detection of heart diseases using fuzzy logic. *International Journal of Engineering Trends and Technology (IJETT), 4*(6), 2694–2699.

30. Sagi, A., Sabo, A., Kuljić, B., & Szakáll, T. (2010, November). Neuro-fuzzy systems in medicine. In *2010 11th International Symposium on Computational Intelligence and Informatics (CINTI)* (pp. 293–296). IEEE.

31. Oad, K. K., DeZhi, X., & Butt, P. K. (2014). A fuzzy rule based approach to predict risk level of heart disease. *Global Journal of Computer Science and Technology, 14*(3-C), 2694–2699.

32. Valarmathi, S., Sulthana, A., Rathan, K. R., Balasubramanian, C. L., & Sridhar, R. (2012). Prediction of risk in breast cancer using fuzzy logic tool box in MATLAB environment. *International Journal of Current Research, 4*(09), 072–079.

33. Basha, S. S., & Prasad, K. S. (2008). Automatic detection of hard exudates in diabetic retinopathy using morphological segmentation and fuzzy logic. *International Journal of Computer Science and Network Security, 8*(12), 211–218.

5 Deep Learning Approach to Predict and Grade Glaucoma from Fundus Images through Constitutional Neural Networks

Kishore Balasubramanian
Dr Mahalingam College of Engineering and Technology

N. B. Ananthamoorthy
Hindusthan College of Engineering and Technology

CONTENTS

5.1 INTRODUCTION

Glaucoma is a neurodegenerative disease that progressively damages the optic nerve causing irreversible complete or partial loss of vision. Intraocular pressure (IOP) in the eye gets increased [1]. The reason for the disease is unknown but can affect anyone irrespective of age, gender, etc. Ophthalmologists control the IOP to protect the subject from going blind. The subject does not recognize the onset of the disease unless some symptoms are shown like narrowing of vision, pains in the eye, etc. [2]. Early diagnosis and screening continuously may aid in preventing the loss of vision. Clinical examination by experts in ophthalmology for glaucoma is done by tonometry, optical coherence tomography (OCT) and Heidelberg retinal tomography (HRT). The methods are very expensive and require skill by experts to detect the disease, which may still be subjective [3]. Glaucoma can be detected by anatomical changes in the eye and often diagnosed by the information from clinical examinations, namely the visual field (VF), OCT, and fundus photo. Glaucoma is characterized by the deterioration of optic nerve fibers (ONF) followed by increased IOP. Normally glaucoma progression is measured from the geometrical parameters of the optic nerve head (ONH). These parameters measure the variations in the ONH structures such as the diameter and area of the optic disk (OD), diameter of the optic cup, man cup depth, and rim area [4]. These changes are easily detected from the fundus images of the ONH and hence are commonly deployed to record the features of disk, cup, and rim [5]. Glaucoma diagnosis is usually conducted by calculating the cup-to-disc ratio (CDR), inferior superior nasal temporal rule (ISNT), glaucoma risk index (GRI), and some more factors from the image. These factors depend upon the extraction and localization of OD and optic cup (OC) [6]. Among these factors CDR is considered to be an important tool to assess glaucoma progression. It is seen from the reports that a CDR value of 0.3 is considered as a normal value. CDR greater than 0.3 is considered an indicating factor of glaucoma [7–8]. Depending on the value of CDR and presence of the superior or inferior focal notches of the retinal image, glaucoma can be classified into the mild, moderate, and severe glaucoma [9]. A CDR in the range of 04–0.7 results in mild glaucoma where side, or peripheral, vision loss is encountered. CDR greater than 0.7 causes ONH hemorrhage, called moderate glaucoma, and if left untreated leads to the advanced stage called severe glaucoma, where the central vision also gets affected [10]. It is also inferred that the IOP for mild glaucoma ranges from 15 to 17 mmHg, for moderate glaucoma from 12 to 15 mmHg, and for severe glaucoma it is between 10 and 12 mmHg [11]. As glaucoma is related to ocular pressure, it is important to know about ocular hypertension (OHT). In OHT, IOP seems to be higher than normal but neither damages optic nerves nor causes loss of vision. It must be monitored closely, as it is considered to be a glaucoma risk factor. From the literatures it is seen that the diagnosis of glaucoma employing clinical trials is extremely tedious, laborious, and can cause human observer variability errors. This affects the accuracy and outcome of the investigation. Figure 5.1 shows sample normal and glaucomatous image. Figure 5.2 shows the stages of glaucoma with OHT.

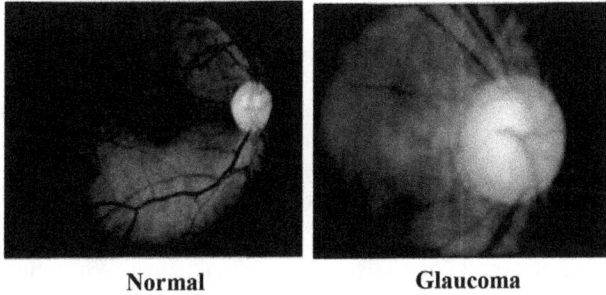

Normal **Glaucoma**

FIGURE 5.1   Anatomy of a sample normal and glaucoma eye.

Industrial revolution and its recent advents in the use of artificial intelligence (AI) for medical diagnosis have allowed us to design and develop non-invasive computational tools by the way of computer-aided diagnosis (CAD) to detect glaucoma in its early or beginning stage using retinal fundus images. The system can act as a second opinion to the ophthalmology specialist aiming at improving the accuracy and precision in the diagnosis [12]. The system offers computational speed and offers a solution to complex problems combining the human expert intelligence and artificial intelligence. In order to achieve optimal performance, the system has to be developed to handle a much larger database [13–14]. CAD systems are designed based on programs in the area of a particular knowledge domain, namely medical systems for continuous support and aid in decision making. The system development has increased manyfold due to their easy implementation on any well-defined medical dataset. These systems are called expert systems in medical diagnosis [15]. CAD systems are fast, cost effective, and could be employed even without the help of an ophthalmic expert, thus avoiding observer variability error. It is recommended by the American Academy of Ophthalmology to have a routine screening for glaucoma once in every one or two years after the age of 65 and every two to four years for people in the age group 40–65 [16]. Diagnosing glaucoma at beginning or early stages will save vision loss [17]. In this chapter, glaucoma diagnosis through deep

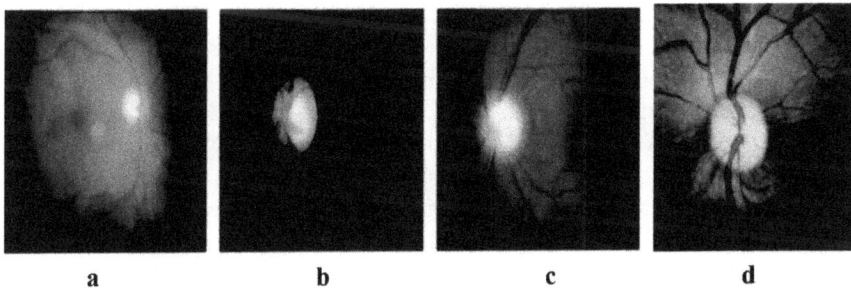

a   b   c   d

FIGURE 5.2   Stages of glaucoma: (a) mild, (b) moderate, (c) severe, and (d) OHT.

learning method is discussed. The structured convolutional neural network (CNN) is employed to effectively grade glaucoma based on the severity level and the novelty is that the developed structure can also identify the suspect class called OHT, which serves as the risk factor for glaucoma occurrence. There is enough evidence that a thin central cornea, high baseline IOP, and increased vertical CDR are significant risk factors that contribute individually for glaucoma progression. CNN are applied in numerous computer vision tasks namely medical image classification [18], pattern recognition, etc. Previous studies have shown hand-crafted machine learning features that depend on expert knowledge and are often tedious. When these are applied to large datasets, their representative power is lost and enhanced classification and feature extraction is required. This is possible by using deep learning architectures.

The approach addresses the following:

1. Minute features in the retinal images are captured
2. Size of the image used fits in the CNN as bigger samples increase the computation time.
3. Different kernel sizes are employed to capture very small changes in the image.
4. Deep networks increase the overall performance by giving consistent performance for most of the iterations.
5. Number of epochs, large datasets, and training parameters used reduces the problem of over fitting.

Organization of the chapter is as follows: Section 5.2 presents literatures studies in the field of study, Section 5.3 outlines the materials and methods adopted, outcomes of the proposed system is discussed in Section 5.4, and Section 5.5 provides the Conclusion

5.2   RELATED WORKS

This section presents an overview of some of the techniques used to detect glaucoma using image processing and computer vision approaches. Automated diagnosis of glaucoma is a challenging and categorical task due to the geometrical features of the eye and its surrounding areas. Most research works relating to glaucoma rely on CDR for assessing the disease. Using end-to-end CNN for classification is another method. CNNs employ 2D kernels, but in recent publications, 3D kernels were also deployed to segment the images [19]. There are four basic operations in CNNs: convolution, non-linearity, pooling or sub-sampling, and classification, which act on the images in matrix form with pixels [20].

In 2015, Chen, Xu, et al, proposed a method involving CNN that captures the features to discriminate glaucoma-related hidden patterns [21]. The deep learning (DL) architecture proposed had six learned layers: four convolutional layers and two fully-connected layers. Softmax classifier was used for predicting glaucoma, which gets input from the last fully connected layer. Experimental results on the two databases ORIGA and SCES achieved much better results than the existing algorithms.

Orlando, Prokofyeva et al, proposed in 2017 a novel method for detecting glaucoma using pre-trained CNN from non-medical data [22]. The method employed OverFeat and VGG-S, two different CNNs, which were applied to retinal fundus images resulting in the generation of feature vectors. Certain preprocessing techniques were employed to improve feature discrimination. The technique was tested on the Drishti-GS1 database and the evaluation was done on average ROC curve. The technique yielded sound results and made the approach viable for learning.

In 2017, Selvastopolsky [18] modified U-Net CNN and proposed to segment OD and OC using DL. Better computational time was recorded on DRIONS-DB, RIM-ONE v.3, DRISHTI-GS datasets. Quality of OD segmentation surpassed other methods while OC segmentation was still a challenging task. 0.95 F-Score for OD segmentation and 0.85 for OC segmentation were achieved.

Ferreira et al, in 2018 proposed a method for the diagnosis of glaucoma using CNN. OD segmentation is performed using CNN and texture descriptors were employed based on phylogenetic analysis to characterize the ROIs. The method was tested on RIM-ONE, DRIONS-DB, and DRISHTI-GS datasets. Results obtained were promising, reaching 100% on all metrics in the red channel analysis as reported in [23].

In the work proposed by Benzebouchi et al, in 2017, a CNN approach was presented to diagnose glaucoma using multimodal data from fundus images. High classification accuracy was reported when tested on RIM –ONE data set. A novel feature was that the classical features extraction step was avoided by processing the feature extraction step and one time classification within the same network of neurons to eventually obtain an automated diagnosis in the absence of user input [24].

In 2017, Miikkulainen et al emphasized that CNN's success depended on identifying the best architecture for the task [25]. In the literature it was stated that the architectures were evolved after lot of trials with complex topologies and a number of parameters resulting in unexpected results. In this work, genetic algorithms (GAs) have been employed successfully to sequential decision tasks. Examples of good CNN applications combined with GAs were reported in Xie and Yuille 2017, Sun et al 2017 [26, 27].

As seen from the literature, different machine learning algorithms (MLA) were proposed and compared to study their performance for diagnosing glaucoma. Some of them were multilayer perceptron networks (MLP), support vector machines (SVM), and feed forward propagation (FFP) among the other algorithms. A high degree of performance was reported during the classification phase [28, 29]. Even though various approaches based on neural networks and classical classifiers were producing good results, recent advents in computational vision have advertised the deployment of CNNs for pattern recognition in retinal images [30]. A survey of various segmentation procedures for OD and OC and their performance comparison, which aid in glaucoma detection, are available in the literature as seen in [31–33]. Most recently, tensor flow [34] has been widely used to construct CNNs. Tensor flow is specialized open source software for text, speech, and image recognition developed by Google. It offers great flexibility in CNN construction and training.

5.3   MATERIALS AND METHOD

5.3.1   DATABASE AND DIVISION OF IMAGES

In this section, the deployment of the proposed method, feature extraction, and classification to diagnose glaucoma and its stages is presented. The first step is the acquisition of digital fundus images from the public databases like RIM-ONE, DRIONS-DB, DRIVE, HRF, and DRISHTI – GS. Certain images are picked up for training the CNN. For this work, a large number of datasets was collected to analyze and investigate the onset of glaucoma severity level through the DL approach. The datasets are labeled by specialists in the field and certain ground truth images were also considered. DL approach was applied on the databases to assess the viability of the technology via performance on images with varying resolutions [35]. The data acquisition resulted in 1155 fundus images collected through fundus cameras, namely Canon, Nidek AFC-210, and Top-con. Images had varying resolutions, for example 1924 × 1556 pixels and 2336 × 2336 pixels. Each image class sample was classified randomly into entirely non-dependent sub-groups for the purpose of training and testing. The testing class image is invariably same for all the glaucoma grades in such way to reduce potential bias occurring due to sample size imbalance. The images were rated by experienced professionals in grading the images into different classes. Table 5.1 presents the division of acquired fundus images based on their severity level.

5.3.2   PREPROCESSING AND DOWN-SAMPLING

To remove noise and unwanted backgrounds from the acquired images, the image preprocessing step is employed. Certain morphological operations like contrast enhancement, binarization, erosion, and dilation are performed to preserve the field of view (FOV) of the images [36]. Over fitting due to the small number of datasets can be avoided by data augmentation processes. The data augmentation increases the variation within a training dataset. This repetition process involves cropping, random translations, and flipping the images to a fixed downscaled size of 227 × 227 [37].

TABLE 5.1
Division of Images for Glaucoma Identification Based on Severity Level

Stages	Healthy	Glaucoma		
		Early	Moderate	Deep
Training	301	142	84	228
Testing	100	100	100	100
Total	401	242	184	328

5.3.3   DEEP LEARNING ARCHITECTURE (DLA)

A DL architectural framework is an integration of feature extraction and classification.

The strength of DLA depends on weight sharing. This is constructed with an assumption that the image may contain similar structures in various locations. Convolution filters can detect these structures with kernel functions. Non-linear activation transform, pooling function, and translation invariance are the operations that take place in the CNNs [38]. A CNN is a DL model that comprises of the following network layers: input layer, convolutional layer, batch normalization layer, pooling layer, softmax layer, etc. The size of the input decides the number of layers; not necessarily all the layers have to be used for the application. The deeper the network the better is the training, but at the same time computational time increases. One important property of CNN is that they have the ability to self-learn and self-organize unsupervised [39]. In order to obtain better training and reduce computational time, one can use the minimum number of layers to get the maximum benefit by employing efficient network parameters. By tuning the network parameters effectively it is possible to achieve a better classification rate. A brief note on the layers of CNN is given below

5.3.3.1   Input Layer

It must be present in the CNN to take the input forward to the network.2D and 3D forms of the images are accepted by the layer with dimensions being initialized.

5.3.3.2   Convolutional Layer

Convolution of the input image is performed in this layer resulting in feature maps that are used as input to the next layer.

5.3.3.3   Batch Normalization Layer

This layer ensures fast learning and allows the flow of normalized samples in between the intermediate layers, improving overall performance.

5.3.3.4   Rectified Linear Unit

This layer reduces data redundancy without losing vital information.

5.3.3.5   Max Pooling Layer

Applying the max pooling operation on each feature map reduces the size of the map according to the user-chosen value.

5.3.3.6   Fully Connected Layer

The max pooling layer donates neurons to this layer and they are connected to every neuron of this layer. The number of classification classes depends on the output of this layer.

5.3.3.7   Softmax Layer

A normalized exponential function operation called the softmax is performed which reduces the data sample dimensionality by removing the outliers so that they fall in the range 0–1. The output label is determined by this layer.

5.3.4 Methodology

5.3.4.1 Proposed CNN Architecture

CNNs have two parts, the convolutional layer and the fully connected layer. In the convolutional layer the input of every layer serves as the output of the preceding layer that acts as feature extractor. These extracted features are classified in the fully connected layer followed by the softmax activation function. Each class has only one output neuron. Note that by tuning the parameters misclassification error is reduced. The overview of the 25-layer CNN is shown in the Figure 5.3 and Table 5.2 presents the parameters used to construct the CNN.

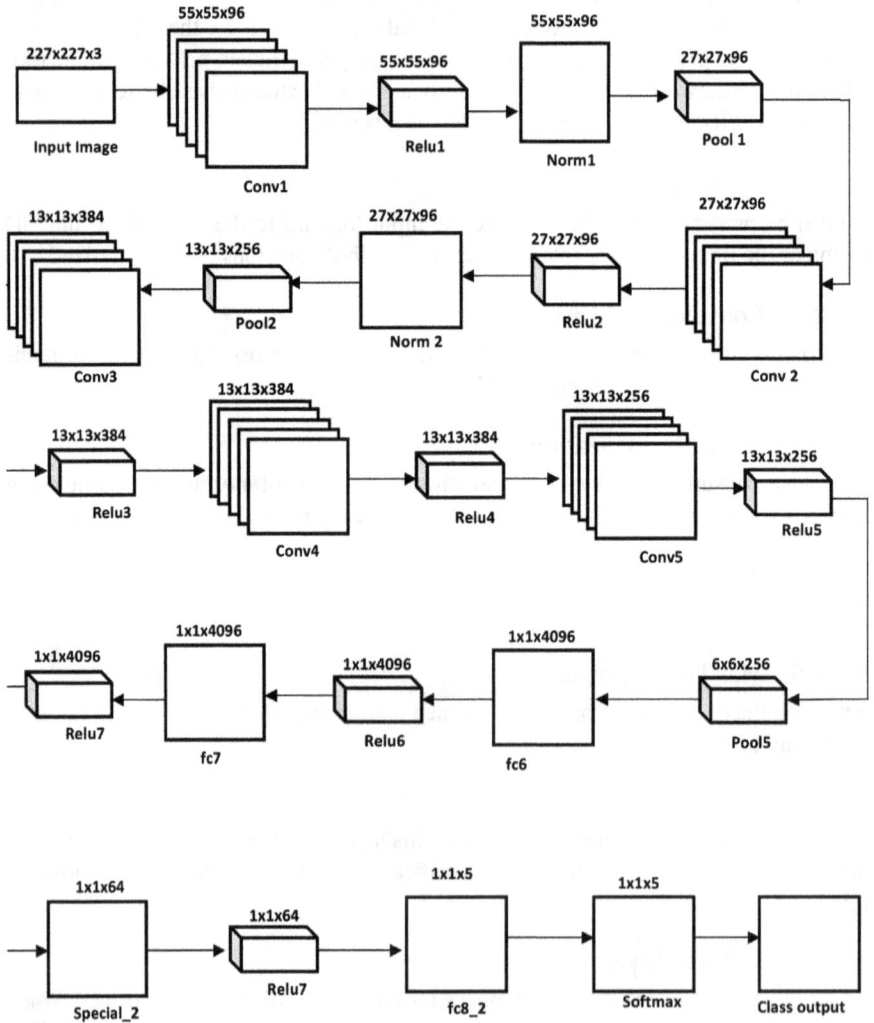

FIGURE 5.3 Overview of proposed CNN architecture.

TABLE 5.2
Parameter Description of the Proposed CNN

S. No	Name	Type	Activations
1	Image input $227 \times 227 \times 3$ image with 'zero center normalization'	Image Input	$227 \times 227 \times 3$
2	conv1-96 $11 \times 11 \times 3$ convolution with stride [4 4] and padding [0 0 0 0]	Convolution	$55 \times 55 \times 96$
3	relu 1-ReLU	ReLU	$55 \times 55 \times 96$
4	norm 1-cross channel normalization with 5 channels per element	Cross Channel Normalization	$55 \times 55 \times 96$
5	pool 1 3×3 max pooling with stride [2 2] and padding [0 0 0 0]	Max Pooling	$27 \times 27 \times 96$
6	conv2-256 $5 \times 5 \times 48$ convolution with stride [1 1] and padding [2 2 2 2]	Convolution	$27 \times 27 \times 96$
7	relu2 -ReLU	ReLU	$27 \times 27 \times 96$
8	norm2-Cross channel normalization with 5 channels per element	Cross Channel Normalization	$27 \times 27 \times 96$
9	pool2 3×3 max pooling with stride [2 2] and padding [0 0 0 0]	Max Pooling	$13 \times 13 \times 256$
10	conv3-384 $3 \times 3 \times 256$ convolutions with stride [1 1] and padding [1 1 1 1]	Convolution	$13 \times 13 \times 384$
11	relu3-ReLU	ReLU	$13 \times 13 \times 384$
12	conv4 - 384 $3 \times 3 \times 192$ convolutions with stride [1 1] and padding [1 1 1 1]	Convolution	$13 \times 13 \times 384$
13	relu4-ReLU	ReLU	$13 \times 13 \times 384$
14	conv3-256 $3 \times 3 \times 192$ convolutions with stride [1 1] and padding [1 1 1 1]	Convolution	$13 \times 13 \times 256$
15	relu5-ReLU	ReLU	$13 \times 13 \times 256$
16	pool5 3×3 max pooling with stride [2 2] and padding [0 0 0 0]	Max Pooling	$6 \times 6 \times 256$
17	fc6 4096 fully connected layer	Fully Connected	$1 \times 1 \times 4096$
18	relu6-ReLU	ReLU	$1 \times 1 \times 4096$
19	fc7 -4096 fully connected layer	Fully Connected	$1 \times 1 \times 4096$
20	relu7-ReLU	ReLU	$1 \times 1 \times 4096$
21	special_2 64 fully connected layer	Fully Connected	$1 \times 1 \times 64$
22	Relu-ReLU	ReLU	$1 \times 1 \times 64$
23	fc8_2 -5 fully connected layer	Fully Connected	$1 \times 1 \times 5$
24	Softmax -Softmax	Softmax	$1 \times 1 \times 5$
25	Class output cross entropyex	Classification Output	-

5.3.4.2 Training and Testing Schemes

The developed CNN architecture is trained to classify images into normal and glaucoma and, further, glaucomatous cases in terms of severity using the local and public image sub-groups. From the acquired dataset, the training datasets are provided as inputs to the network. The images are then preprocessed, augmented, downsized, and fed to the CNN. One hundred images are selected for testing. The rest are selected for training the network. Many iterations are carried out randomly to generalize the performance. Stochastic gradient descent momentum training, also known as the steep descent, with a batch size of 30 samples is used to reduce the entropy loss. The momentum set is 0.9. The system is iterated with learning rates of 0.1, 0.01, 0.001, 0.0001, and 0.00001 with 300 epochs. The training rate must be optimal, as too low or high may cause long computational time or error respectively. Data augmentation is employed to prevent the problem of overfitting with the aim to artificially enlarge the training data set. Horizontal and vertical flipping is carried out randomly in training and image rotation is done via +30° to −30° [40].

5.4 EXPERIMENTS

5.4.1 Performance Assessment

All the retinal fundus images are resized to 227×227 and subjected to the developed CNN. Experiments are performed on MATLAB, Intel Core i7 processor, 16GB RAM with GPU. The evaluation performance was quantitatively assessed using the metrics shown in Table 5.3

TABLE 5.3
Performance Parameters Used for the Developed CNN Model

Parameter	Expression
Sensitivity	$\dfrac{TP}{TP+FN}$
Specificity	$\dfrac{TN}{TN+FP}$
Accuracy	$\dfrac{TP+TN}{TP+FN+TN+FP}$
Precision / Positive Predictive Value	$\dfrac{TP}{TP+FP}$
Recall	$\dfrac{TP}{TP+FN}$
F-Score	$2\times\dfrac{(Precision)(Recall)}{Precision+Recall}$

TABLE 5.4

Results of the Developed CNN for 227×227 Input Size

Learning Rate	Sensitivity	Specificity	Accuracy	Precision	F-Score
0.1	0.95	0.966	0.94	0.95	0.84
0.01	0.9	0.941	0.956	0.91	0.88
0.001	**0.98**	**0.984**	**0.989**	**1**	**1**
0.0001	0.89	0.924	0.923	0.8	0.842
0.00001	0.797	0.832	0.813	0.9	0.857

Upon generalization of the results, the experiment is randomly iterated for 50 times. The average of all performance metrics for all the 50 iterations are computed and recorded. Tables 5.4 and 5.5 presents the results for the developed CNN.

The CNN can also classify the glaucomatous images into mild, moderate, and severe glaucoma. The OHT class is also added, which will serve as an indicator for glaucoma progression. Table 5.6 shows the performance of the CNN in terms of accuracy in discriminating the glaucomatous cases into different grades. Figure 5.4 Shows accuracy plot for various learning rates in grading glaucoma severity level.

5.5 DISCUSSION AND CONCLUSION

A CNN can have many hyper-parameters which may require further tuning, including number of fully connected layers, convolutional layers, pooling layers, filters, hidden nodes, learning rate, etc. Construction of a CNN can be time consuming if it is designed from scratch and also expensive computationally. Transfer learning could help to solve the problem by leveraging the features trained by a pre-training DL model and then applying it to various datasets. A CNN provides better performance if there is a greater number of datasets [41]. From the results, except for the learning rate 0.001, all the others have a comparatively low performance rate in terms of accuracy. The network might have missed some subtle information that serves to be vital when trained with those learning rates. Image

TABLE 5.5

Average Performance of the CNN Model with 50 Iterations

Learning Rate	Sensitivity	Specificity	Accuracy	Precision	F-Score
0.1	0.956	0.932	0.921	0.94	0.84
0.01	0.965	0.949	0.938	0.9	0.852
0.001	**0.949**	**0.951**	**0.9765**	**0.9**	**1**
0.0001	0.872	0.919	0.88	0.75	0.857
0.00001	0.791	0.832	0.833	0.88	0.81

TABLE 5.6
Performance of CNN in Grading Glaucoma in Terms of Accuracy

Learning Rate	Accuracy			
	Early	Moderate	Deep	OHT
0.00001	0.65	0.6	0.58	0.75
0.0001	0.9	0.85	0.85	0.9
0.001	**0.977**	**0.96**	**0.98**	**0.98**
0.01	0.88	0.87	0.77	0.65

size also could play an important factor. Larger image size may require more computational time. To get a compromise between the two, 227×227 is chosen in the model. The model works well with different kernel sizes to attract even minor changes. Approximately 99% specificity is also achieved, making the system robust to almost all normal and abnormal cases. The experiment is iterated and the average of all the performance metrics is computed. Preprocessing is done to preserve the FOV of the images since the system can be used to further grade the disease. Table 5.7 shows the summary of and comparison of various DL methods developed for detecting glaucoma. One of the main merits of using CNN is that it does not require the normal steps of feature extraction, ranking, dimensionality reduction, etc. Feature maps are generated by extracting features automatically after each layer and taking the best features based on self-decision. It is seen that design and development of such a tool will definitely help ophthalmologist to diagnose glaucoma even at the onset of the disease or when analyzing the risk factors associated with it. The method developed is able to achieve an accuracy of

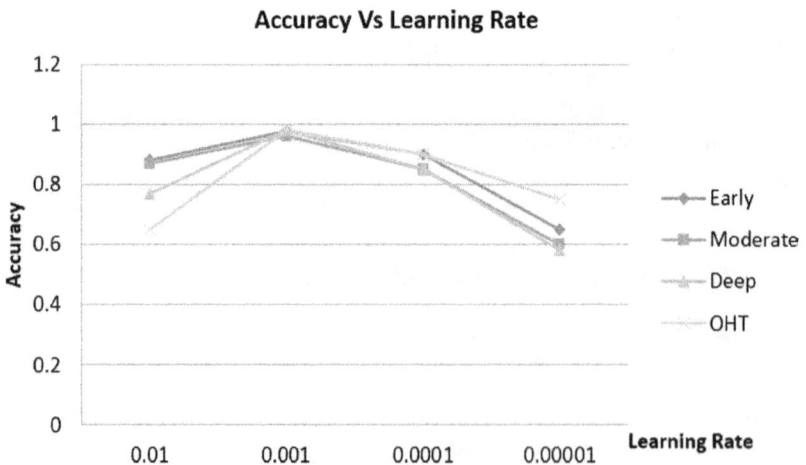

FIGURE 5.4 Accuracy vs. learning rate for glaucoma stages.

TABLE 5.7
Summary and Comparison of Various DL Methods Developed for Detecting Glaucoma

References	Method	Datasets	Performance Indices
Al-Bander et al [43]	23 layer CNN, SVM	RIM-ONE	Accuracy – 85%, Sensitivity – 85%, Specificity – 89.8%
Fu et al [44]	Ensemble of 4 CNN	ORIGA, SCES	Accuracy – 91.83%, Sensitivity – 84.8%, Specificity – 83.8%
Chen et al [21]	6 layer CNN	ORIGA, SCES	AUC – 83% to 88%
Shibata et al [45]	Transfer Learning with ResNet	Private	AUC – 96.5%
N E Benzebouchi et al [24]	Cooperative CNN	RIM-ONE	Accuracy – 96.9%, Sensitivity – 96.5%, Specificity – 97.3%
Proposed CNN	25 layer CNN	5 Public dataset, 1155 images	Accuracy – 98.9%, Sensitivity – 98%, Specificity – 98.4%, Precision and F-Score – 100%

98.9% with 98% sensitivity and specificity. The CNN can detect the normal class and glaucoma class and also can grade the images into classes: early, moderate, and deep. The OHT class is also included to assess the risk factor. When DL is used, original fundus image resolution can be reduced, taking into account the limitations of computer memory and hardware. Though there is a considerable amount of dataset, the DL method warrants much more dataset. It can be seen from [42] that a greater number of datasets is used in the study. From Table 5.2 the categorization of images may not be completely perfect due to potential bias. The developed CNN model can aid in early diagnosis of glaucoma and also predict the suspect class with a high accuracy. In future, the system can be developed with multiscale architectures to grade glaucoma with a greater number of images so that the computational time gets reduced further.

CONFLICT OF INTEREST

The authors disclose no potential conflicts of interest and funding.

REFERENCES

1. A. Sommer, J.M. Tielsch, J. Katz, H.A. Quigley, J.D. Gottsch, J. Javitt, K. Singh, "Relationship between intraocular pressure and primary open angle glaucoma among white and black Americans", The Baltimore Eye Survey, Archives of Ophthalmology 109(8) (1991), pp. 1090–1095.

2. J. Nayak, U.R. Acharya, P.S. Bhat, A. Shetty, T.C. Lim, "Automated diagnosis of glaucoma using digital fundus images", Journal of Medical Systems 33(5) (2009), pp. 337–346.

3. P.S Mittapalli, G.B. Kande, "Segmentation of optic disc and optic cup from digital fundus images for the assessment of glaucoma", Biomedical Signal Processing and Control 24 (2016), pp. 24–36.

4. S. Miglior, M. Guareschi, E. Albe, S. Gomarasca, M. Vavassori, N. Orzalesi, "Detection of glaucomatous visual field changes using the Moorfields regression analysis of the Heidelberg retina tomography", American Journal of Ophthalmology 136(1) (2003), pp. 26–33.

5. P. Betz, F. Camps, J. Collignon-Brach, G. Lavergne, R. Weekers, "Biometric study of the disc cup in open-angle glaucoma", Graefe's Archive for Clinical and Experimental Ophthalmology 218 (2) (1982), pp. 70–74.

6. R. Chrástek et al, "Automated segmentation of the optic nerve head for diagnosis of glaucoma", Medical Image Analysis 9(4) (2005), pp. 297–314.

7. The Eye Digest, "Eye exam and tests for glaucoma diagnosis", The University of Illinois Eye and Ear Infirmary, Archived 2012.

8. Garway-Heath et al, "Vertical cup-disc ratio in relation to optic disc size: its value in the assessment of glaucoma suspect", British Journal of Ophthalmology 82(10) (1998), pp. 1118–1124.

9. Y. Lakshmanan, R.J. George, "Stereo acuity in mild, moderate and severe glaucoma", Ophthalmic and Physiological Optics 33(2) (2013), pp. 172–178.

10. A.J. Vingrys, "The many faces of glaucomatous optic neuropathy", Clinical and Experimental Optometry, 83 (2000), pp. 145–160.

11. R. Sihota, D. Angmo, D. Ramaswamy, T. Dada, "Simplifying 'target' intraocular pressure for different stages of primary open-angle glaucoma and primary angle-closure glaucoma", Indian Journal of Ophthalmology 66 (2018), pp. 495–505.

12. M.L. Giger, "Computer-aided diagnosis", RSNA Categorical Course in Physics (1993), pp. 283–298.

13. H. Takahashi et al, "Applying artificial intelligence to disease staging: Deep learning for improved staging of diabetic retinopathy", PLoS One 12(6) (2017).

14. V.B. Ciesielski et al, "Applications-oriented AI research: Medicine technical report", Stanford University CA, Department of Computer Science (1979).

15. C.A. Kulikowski, S.M. Weiss, "Representation of expert knowledge for consultation: CASNET and EXPERT Projects", In P. Szolovits (Ed), Artificial Intelligence in Medicine, Chapter 2. Boulder, CO: Westview Press (1982).

16. http://www.glaucoma.org/glaucoma/diagnostic-tests.php.

17. D. Kourkoutas et al, "Glaucoma risk assessment using a non-linear multivariable regression method", Computer Methods and Programs in Biomedicine 108(3) (2012), pp. 1149–1159.

18. A. Sevastopolsky, "Optic disc and cup segmentation methods for glaucoma detection with modification of U-Net convolutional neural networks", Pattern Recognition and Image Analysis 27(3) (2017).

19. B. Kayalibay et al, "CNN-based segmentation of medical imaging data." CoRRabs/1701.03056 (2017).

20. A. Geitgey, "Deep learning and convolutional neural networks," Jun 13, 2016. [Online] Available: https://medium.com/@ageitgey/machine-learning

21. Xiangyu Chen et al, "Glaucoma detection based on deep convolutional neural network", Proceedings of the 37th Annual International Conference of the IEEE Engineering in Medicine and Biology Society (EMBC), IEEE (2015), pp. 715–718.

22. J. Orlando, E. Prokofyeva, M. del Fresno, M.B. Blaschko, "Convolutional neural network transfer for automated glaucoma identification", Proc. SPIE 10160, 12th International Symposium on Medical Information Processing and Analysis, 101600U (2017).
23. M.V. Ferreiraa et al, "Convolutional neural network and texture descriptor-based automatic detection and diagnosis of glaucoma", Expert Systems with Applications 110(15) (2018), pp. 250–263.
24. N.E. Benzebouchi, N. Azizi, S.E. Bouziane, "Glaucoma diagnosis using cooperative convolutional neural networks", Proceedings of ISER 88th International Conference, Venice, Italy (2017).
25. R. Miikkulainen, et al, "Evolving deep neural networks", CoRR, abs/1703.00548 (2017).
26. Sun, Y., Xue, B., and Zhang, M. "Evolving deep convolutional neural networks for image classification", CoRR, abs/1710.10741 (2017).
27. Xie, L. and Yuille, A.L., "Genetic CNN", CoRR, abs/1703.01513 (2017).
28. Kwokleung Chan, Te-Won Lee, P. A. Sample, M. Goldbaum, R. N. Weinreb, T. J. Sejnowski, "Comparison of machine learning and traditional classifiers in glaucoma diagnosis", IEEE Transactions on Biomedical Engineering 49(9) (2002). doi: 10.1109/TBME.2002.802012.
29. O. Sheeba, "Glaucoma detection using artificial neural network", IACSIT, Ed. IACSIT, 6(2) (2014)
30. H. Eckerson, "Deep learning past, present, and future," 2017. [Online] Available: https://www.kdnuggets.com/2017/05/deep-learning-big-deal.html
31. A. Allam, A. Youssif, A. Ghalwash, "Automatic segmentation of optic disc in eye fundus images: A survey, ELCVIA: Electron", Electronic Letters on Computer Vision and Image Analysis 14 (2015), pp. 1–20.
32. A. Almazroa et al, "Optic disc and optic cup segmentation methodologies for glaucoma image detection: a survey", Journal of Ophthalmology 2015 (2015), p. 28.
33. N. Thakur, M. Juneja, "Comparative analysis on optic cup and disc segmentation for glaucoma diagnosis", Springer Proceedings of 3rd International Conference on Advanced Computing, Networking and Informatics (2016), pp. 219–223.
34. D. Calkins, "Glaucoma", Tech. Rep. 88-207450, 2016. [Online] Available: https://www.glaucoma.org/GRF_Understanding_Glaucoma_ES.pdf
35. K. He, X. Zhang, S. Ren, J. Sun, "Deep residual learning for image recognition", In CVPR (2016).
36. S. Mukhopadhyay et al, "Multiscale morphological segmentation of gray scale images", IEEE Transactions on Image Processing 12 (2003), pp. 533–549.
37. S.C. Wong, A. Gatt, V. Stamatescu, M.D. McDonnell, "Understanding data augmentation for classification: When to warp?" International Conference on Digital Image Computing Tech Appl DICTA 2016 (2016). https://doi.org/10.1109/DICTA.2016.7797091
38. G. Litjens, T. Kooi, B. Bejnordi, A. Setio, F. Ciompi, M. Ghafoorian, J. Laak, B. Ginneken, C. Sanchez, "A survey on deep learning in medical image analysis", Medical Image Analysis 42 (2017), pp. 60–68.
39. U. Rajendra Acharya et al, "Automated detection of arrhythmias using different intervals of tachycardia ECG segments with convolutional neural network", Information Sciences, 405 (2017), pp. 81–90.
40. L. Wang et al, "2D/3D rigid registration by integrating intensity distance", Optics and Precision Engineering, 22 (2014), pp. 2815–2824.
41. U Rajendra Acharya et al, "Automated identification of shockable and non- shockable life threatening ventricular arrhythmias using convolutional neural networks", Future Generation Computer Systems 79 (2017), 952–959.

42. A. Krizhevsky, I. Sutskever, G. E. Hinton, "ImageNet classification with deep convolutional neural networks", Communications of ACM 60(6) (2017), pp. 84–90.

43. B. Al-Bander, W. Al-Nuaimy, M.A. Al-Taee, Y. Zheng, "Automated glaucoma diagnosis using deep learning approach", Proceedings of 14th International Multi-Conference on Systems, Signals Devices (SSD) (2017), pp. 207–210.

44. H. Fu, J. Cheng, Y. Xu, C. Zhang, D.W.K. Wong, J. Liu, and X. Cao, "Disc-Aware ensemble network for glaucoma screening from fundus image", IEEE Transactions on Medical Imaging 37(11) (2018), 2493–2501.

45. N. Shibata, M. Tanito, K. Mitsuhashi, Y. Fujino, M. Matsuura, H. Murata, and R. Asaoka, "Development of a deep residual learning algorithm to screen for glaucoma from fundus photography", Scientific Reports 8(1) (2018), 14665.

6 A Novel Method for Securing Cognitive Radio Communication Network Using the Machine Learning Schemes and a Rule Based Approaches

Antony Hyils Sharon Magdalene
and Lakshmanan Thulasimani
PSG College of Technology

CONTENTS

6.1   INTRODUCTION

Cognitive radio (CR) is an important solution to the problem of wireless spectrum collision. The frequency white space problem is encountered in the 5G period [1]. For the growth of 5G and the Internet of Things (IoT), security is essential for the next generation mobile network [2]. A cognitive radio network (CRN) contains licensed primary users (PUs) and an unlicensed secondary users (SUs) [3]. The CRN discovers the vacant channel and the SU retransmits on the available channel. If the SU detects any PU signal utilizing the channel, then the SU moves backward and searches for another channel [3].

Cooperative spectrum sensing (CSS) coordinates and sends a decision to the fusion center (FC) that will execute the final decision. It uses spatial and multiuser diversity by raising the probability of detection and by reducing the probability of false alarm. This is the process for a vacant spectrum [4].

In CSS, assailant sensors are present in the cooperative model. This misleads the FC to make an incorrect decision. The CSS [5] is harmed by the malicious counterfeit, which is known as a spectrum sensing report forge (SSRF) [6] attack or a spectrum sensing data falsification (SSDF) attack. A comprehensive survey of the work in this area is found in [7]. The SSDF attack reports a wrong sensing result to the FC by giving an incorrect resolution. This worsens the total functions of the spectrum sensing phase. In this chapter, the proposed performance of secured CSS is compared with the existing defense system under a fault-barren channel.

The process of CSS requires large communication resources for reporting sensing results. On this account, the performance of hard-decision CSS is studied based on a K-out-of-N rule or counting rule [8–10]. This results in fewer samples for decision making to avoid using too much of the communication resources. It can make global decisions without any prior knowledge of the PU signal.

The suggested scheme uses two cognitive processes, namely, analysis and accessing process. An 'improved-apriori' scheme has been used before in the spectrum sensing process. The suggested scheme achieves better performance for the assailant sensors. For wide-range connectivity, the 5G users approach terminal switches employing various technologies [11]. The CRN concept uses a released spectrum and a high carrier frequency [11]. 5G-based CR shows very high performance in system capacity and data rates with ultra-high reliability and availability, very low latency, and low device cost for energy consumption [12]. This scheme solves SSDF attacks in CR-based 5G for all these categories. It reduces energy usage by identifying the possible availability of channels. This enlarges the data rates by decreasing the error probability. Latency is clearly reduced in the 5G based CR network.

6.2   RELATED WORKS AND EXISTING METHODS

In reference [13], the sequential probability ratio test (SPRT) is proposed to counter the SSDF attack. In [14], a cryptography-based SSDF counter mechanism is proposed for error-free channels. The authors concluded that this approach is efficient under a noisy channel, but has restrictions in the harsh wireless environment. In [15], Lu et al, proposes a hard decision scheme that consists of two tasks: one is an identifying phase and the other is a sensing phase. The first phase identifies authentic SUs in the CRN and the second phase supplies a sensing report to all users to make final decisions. In article [16], the medium access control (MAC) termed sense-and-predict (SaP) is proposed for the CR-based receiver, which faces a different level of an interference according to the separation and beam direction. In SaP, each transmitter predicts the interference level of the receiver. The prediction is based upon the sensed interference at the transmitter. This can be measured by the spatial interference correlation between the two locations. In article [17], a generalized logical sphere (GLS) is proposed for the space-terrestrial networks based on a smart collaborative theory. First, the coverage range, transmission mechanism, and practice guideline of integration development are fully analyzed. Then the advanced features of GLS are presented and discussed. Bhattacharjee et al [18] proposed an apriori algorithm to detect the SSDF attack. This sends a link to every sensing report sent to the FC. An apriori algorithm identifies the link and tries to detect the attack. The adaptive reputation based clustering (ARC) system [19] successfully reduced the error rate and has the ability to analyze the attacking node by outperforming the false detection rates.

In this chapter, the proposed method contains an effective machine learning algorithm called 'improved-apriori' algorithm, that enhances the security of 5G based CR network. Also, the K-out-of-N rule classifies the SUs to perform the task easily. The proposed method works more efficiently than all other existing methods. Analysis show that this technique skillfully secures the CR network from SSDF attack in a blind scenario.

6.3 BRIEF NOTE ON EXISTING METHODS

In all the articles mentioned above, five techniques are implemented as discussed in the following sections.

6.3.1 COLLABORATIVE APPROACH TOWARD SECURE SPECTRUM SENSING

Misbehaving sensing nodes falsify sensing data to prevent legitimate nodes from using the spectrum [20]. A collaborative approach is proposed to monitor behavior and identify malicious nodes. The node's reliability is measured by the belief level (BL).

The clustering method is also used to divide the sensing nodes. The cluster head (CH) is responsible for collecting sensing reports from different sensing nodes to identify the malicious sensing nodes. The CH calculates an adjustment factor (AF) for each reported node and then adds the value of the BL at time varying expression as:

$$AF_{SU_i} = \left(\sum_{g=1,\neq i}^{G} \alpha * N\left(BL_{SU_g}\right) \right) - \left(\sum_{b=1,\neq i}^{B} \beta * N\left(BL_{SU_b}\right) \right) \qquad (6.1)$$

$$s.t. -4 \leq AF \leq 4$$

Where G and B represent the good and bad sensing nodes that decides the SU_i. Let α and β be the rewarding and the penalizing factor and $N(BL_{SU_b})$ the normalized BL of the bad node, that reports SU_i. Then, $N(BL_{SU_g})$ is the normalized BL of the good node, that reports SU_i.

Then, the detection probability of the P_d is calculated as:

$$P_d = \frac{\sum_{b=1,\neq i}^{B} \beta * N(BL_{SU_b})}{\sum_{c=1,\neq i}^{C} \beta * N(BL_{SU_c})} \tag{6.2}$$

Where $N(BL_{SU_c})$ is the normalized value of the BL for each reporting node of SU_i in the same cluster.

6.3.2 SEQUENTIAL 0/1 FOR STRATEGIC BYZANTINE ATTACK

In this chapter [21], a low complexity sequential 0/1 (s0/1) is proposed for CSS in the presence of Byzantine attack. Let P_1 and $\hat{P}_1$ be the probability of the individual report result for the reliable SU and the attacker. The report result is shown as 'spectrum occupancy' and is defined as 1. The average samples that satisfy the voting rule are defined as:

$$\psi\left(N,K,P_1,\hat{P}_1\right) = \sum_{i=1}^{N} \varphi\left(i,K,P_1,\hat{P}_1\right) + \varphi\left(i,N-K+1,1-P_1,1-\hat{P}_1\right) \tag{6.3}$$

The first term in Equation (6.3) has a negative binomial distribution. Let N be the number of SUs. There must be either K report result 1s or $N - K + 1$ report result 0s, that is received at the FC. According to the Bayes theorem, the average samples required for the FC is given by:

$$\bar{N}_{svr} = \psi\left(N,K,P_d,P_{da}|H_1\right)P(H_1) + \psi\left(N,K,P_f,P_{fa}|H_0\right)P(H_0) \tag{6.4}$$

Let H_1 and H_0 be the binary hypothesis, P_f, P_d be the false alarm and the detection probability, and P_{fa}, P_{da} be the false alarm and the detection probability for the attacker. Hence, the average number of samples $s0$ is calculated as:

$$\begin{aligned} \bar{N}_{s0} = &\psi\left(N-K+1,N-K+1,1-P_d,1-P_{da}|H_1\right)P(H_1) \\ &+ \psi\left(N-K+1,N-K+1,1-P_f,1-P_{fa}|H_0\right)P(H_0) \end{aligned} \tag{6.5}$$

For $s1$ the average number is expressed as:

$$\bar{N}_{s1} = \psi\left(N,K,P_d,P_{da}|H_1\right)P(H_1) + \psi\left(N,K,P_f,P_{fa}|H_0\right)P(H_0) \tag{6.6}$$

6.3.3 MALICIOUS COGNITIVE USER IDENTIFICATION ALGORITHM FOR SSDF ATTACK

The method described in reference [22] relies on a β-function and a feedback-iteration method. A multi-channel cooperative condition (β-MIAMC) uses a cognitive user's (CUs) performance rating to analyze the assailant nodes in the multiple sub-channels. β-MIAMC reduces the error accurately and its performance is slightly lower than the proposed method. But the proposed generalized Gaussian distribution (GGD) algorithm outperforms the β-MIAMC method for identifying the channels.

In a β reputation system, the cognitive user send a dichotomy process to a FC with two possible outcomes $\{x,y\}$, where x represents the cognitive user sending a true local result to the FC and y represents the cognitive user sending an SSDF attack. This system is constructed using the following mathematical model:

$$f\left(p_x|\alpha,\beta\right)=\frac{\lceil\alpha+\beta}{\lceil\alpha+\lceil\beta}\,p_x^{\alpha-1}(1-p_x)^{\beta-1} \tag{6.7}$$

Where $p_x \in [0,1]$ and $\lceil(.)$ represents the gamma function.

In the probability density function, the probability $x(p_x)$ is produced by the dichotomy that follows the β distribution. Let r be the number of occurrences of the result x observed and s be the number of occurrences of the result, y. Then the parameters are, $\alpha = r+1$ and $\beta = s+1$, where $r \geq 0$, $s \geq 0$. The expectation of β distribution is:

$$E[p_x]=\frac{\alpha}{\alpha+\beta} \tag{6.8}$$

Where $E[p_x]$ indicates the average probability of the occurrence of result x after the number of occurrences of r and s of the observed result x and y, respectively.

6.3.4 XOR DISTANCE ANALYSIS FOR SECURING CSS

In article [23], the extended design architecture (XDA) scheme is proposed to suppress a SSDF attack. The XDA design contains an XOR distance calculation and a trust mechanism to hold the attack. The calculation is lined up with the historical sensing data of type 0 and 1, which is used to determine the similarity of two SUs. If the XOR distance value is minimized, the SSDF attacker has a high trust value. This performance is much better than the two previous methods. This successfully reduces the power of a SSDF attack.

Using a binary input (0 or 1), the beta function is a popular design to assess a trust value. This computes the amount of positive and negative conducts in which the user has operated. The trust value with the probability of beta density function is counted. This is denoted by Beta(α,β).

$$\text{Beta}(\alpha,\beta)=\frac{\lceil(\alpha+\beta)}{\lceil\alpha\lceil\beta}\,\delta^{\alpha-1}1-\delta^{\beta-1} \tag{6.9}$$

Where δ is the probability of the conducts, which is either positive or negative, $0 \leq \delta \leq 1$, $\alpha > 0$, $\beta > 0$. First get the i-th SU as an example (SU_i). Let hon_i and mal_i

indicate the honest (positive) and the malicious (negative) sensing count, that is conducted by (SU_i). In the baseline, the trust value of SU_i is found as:

$$Tr_i = Beta(hon_i + 1, mal_i + 1) \qquad (6.10)$$

The condition is $\lceil (N) = (N-1)!$, where N is an integer. The expectation value of the beta function is indicated as, $E\left[Beta(\alpha, \beta)\right] = \frac{\alpha}{\alpha+\beta}$. In such case, Tr_i is further calculated as:

$$Tr_i = \frac{1 + hon_i}{2 + hon_i + mal_i} \qquad (6.11)$$

The threshold of trust value is denoted θ. The SU_i is denoted as an attacker, when $Tr_i > \theta$. The threshold θ satisfies two demands. First, it should be a rational value between 0 and 1 as $Tr_i \in [0,1]$. Second, the value of θ is adjusted to reduce the false responses that are generated by the malicious sensing data.

The constituents of the preceding methods (from Sections 6.3.1 to 6.3.4) help to produce computer simulation results. The estimates are based on 5 to 30 attackers.

6.3.5   AN ENHANCED CSS SCHEME BASED ON EVIDENCE THEORY

In article [24], evidence theory is used to mitigate the SSDF attack. First, credibility is calculated in the FC. Based on the credibility, malicious users are reduced by using the weighted probability assignment for each honest user. The collaboration of the attack is determined by evidence theory. In addition, the projection approximation algorithm is also implemented at the FC while processing the local spectrum sensing.

In evidence theory, the frame is defined as, $\{H_1, H_0, \Omega\}$, where Ω denotes the hypothesis is true. In CSS based on evidence theory, each SU needs to assign a basic probability assignment (BPA) to H_1 and H_0, according to the local sensing result. The sample number of $X_i(t)$ is large enough to obey the normal distribution. So, the normal form of membership function is used to get $m_i(H_1)$, $m_i(H_0)$, $m_i(\Omega)$ in two cases of H_1 and H_0, respectively, and is described as:

$$m_i(H_0) = \int_{x_i(t)}^{+\infty} \frac{1}{\sqrt{2\pi\delta_{0i}^2}} \exp\left(-\frac{(x-\mu_{0i})^2}{\delta_{0i}^2}\right) dx \qquad (6.12)$$

$$m_i(H_1) = \int_{-\infty}^{x_i(t)} \frac{1}{\sqrt{2\pi\delta_{1i}^2}} \exp\left(-\frac{(x-\mu_{1i})^2}{\delta_{1i}^2}\right) dx \qquad (6.13)$$

$$m_i(\Omega) = 1 - m_i(H_0) - m_i(H_1) \qquad (6.14)$$

Where μ_{0i}, μ_{1i}, δ_{0i}^2, δ_{1i}^2 denote the mean and variance of $X_i(t)$ under H_0 and H_1, respectively. The above formula shows the credibility of each cognitive user as $m_i(H_0)$ and $m_i(H_1)$.

The above sections, that is from Section 6.3.1 to Section 6.3.5, are implemented as the existing technique and then compared with the proposed results from this chapter's numerical evaluations. The proposed strategy improves the total system performance and exceeds the performance of recent sequential 0/1 techniques.

6.4 SYSTEM MODEL

Table 6.1 shows the symbols and expressions used for the proposed work.

TABLE 6.1
Symbols and Expressions Used for the Proposed Work

Letter	Denotation
S_n	Number of sensors
N_k	Truthful sensor count formulated in the k counter
$\ddot{H}$	Channel real states
$\hat{S}$	Sensor reports
P_f	Probability of false alarm
P_m	Probability of miss-detection
P_{ma}	Malicious sensor probability conducting attacks
P_{fa}	False alarm probability conducting attacks
P_{da}	Detection probability conducting attacks
P_f^T	Truthful sensor's false alarm probability
P_f^{AL}	Assailant sensor's false alarm probability
$\hat{G}$	Universal report
P_d^T	Probability of detection in the truthful sensor
P_d^{AL}	Probability of detection in an assailant sensor
Q_f	Universal probability of a false alarm
Q_m	Universal probability of a miss-detection
P_f^{Ex}	Probability of false alarm for an extended sensing across an entire slit
P_{md}^{Ex}	Probability of miss-detection for an extended sensing across an entire slit
$\dot{R}$	Suggested source results
ρ	Attack strength in percentage
P_{at}	Assailant sensor probability that is incorrectly assigned as a truthful one
P_{ta}	Truthful sensor probability that is incorrectly assigned as an assailant one
P_f^{Pr}	Probability of false alarm for the suggested scheme
P_d^{Pr}	Probability of detection for the suggested scheme
P_{md}^{Pr}	Probability of miss-detection for the suggested scheme
W_f	The universal extension of a particular location for the false alarm probability
W_d	The universal extension of a particular location for the detection probability

Individual sensors operate on energy detection sensing to get a limited decision. An adjacent sensor S_n raises the spectrum sensing possibility. The SU's sensing report is corporate and then a universal decision $\hat{G}$ is constructed for the objective passage. For every sensor, a binary hypothesis test problem is computed in the spectrum sensing as follows [16]:

$$\ddot{H}_0: d(t_0) = w(t_0) \tag{6.15}$$

$$\ddot{H}_1: g.m(t_0) + w(t_0) \tag{6.16}$$

Where $\ddot{H}_0$ refers to PU that is unavailable, $\ddot{H}_1$ denotes that the PU is available, and $d(t_0)$ is the SU's collected signal at time t_0. Let $m(t_0)$ be the PU's transferred signal, g is the profit obtained by the channel, and $w(t_0)$ is the additive white Gaussian (AWG) noise. The collected energy observation in the energy detector is: $E_{ob} = \sum_{i=1}^{2v} |d(t_0)|^2$, where $v = TB$, and where TB is the product of time and bandwidth in the energy detection of CRN. For the central restricted theorem, when v is large enough (e.g., $v \gg 10$), and a Gaussian arbitrary inconstant E_{ob} under two hypotheses $\ddot{H}_0$, then $\ddot{H}_1$ is as follows [18]:

$$\begin{cases} \ddot{H}_0: E_{ob} \sim S(\partial_0, \delta_0^2) \\ \ddot{H}_1: E_{ob} \sim S(\partial_1, \delta_1^2) \end{cases} \tag{6.17}$$

Where $\partial_0 = 2v$, $\delta_0^2 = 4v$, $\partial_1 = 2v(\alpha+1)$, $\delta_1^2 = 4v(2\alpha+1)$, and the SU received signal-to-noise ratio (SNR) is α. By comparing the energy monitoring with a limited threshold φ, the SU binary decision d is given as:

$$E_{ob} \underset{<}{\overset{\geq}{\underset{d_e = 0}{\overset{d_e = 1}{}}}} \varphi \tag{6.18}$$

Let P_0 be the probability of assumptions. The detection probability p_d, and the false alarm probability p_f are formulated as:

$$\begin{cases} P_f^T \triangleq P_0\left(d_e = 1 | \ddot{H}_0\right) = Q\left(\dfrac{\varphi - \partial_0}{\delta_0}\right) \\ P_d^T \triangleq P_0\left(d_e = 1 | \ddot{H}_1\right) = Q\left(\dfrac{\varphi - \partial_1}{\delta_1}\right) \end{cases} \tag{6.19}$$

The Gaussian Q-operation is $Q(x)$. The probability of the missed detection is $P_{md}^T = 1 - P_d^T$. The truthful sensor announces the genuine decision d_e for the data fusion to the SU. Even if an assailant sensor commits wrong decisions f to mislead the SU to construct the falsified conclusion as, $f \neq d_e$. The assailant probability p_a is extended from 0 to 1. Let 0 indicate a peak case of never attack and 1 denote an always-attack case, respectively. Finally, the sensing of assailant sensor is calculated as:

$$\begin{cases} P_f^{AL} = P_f^T.(1 - P_m) + (1 - P_f^T) . P_m \\ P_d^{AL} = P_d^T.(1 - P_m) + (1 - P_d^T) . P_m \end{cases} \qquad (6.20)$$

The SU executes the universal report $\hat{G}$ for report gathering. The common blending rules are K-out-of-N rule and similarity ratio test rule [18]. This is used to execute individual and collaborating sensors. The K-out-of-N rule is suitable for all sensors, but the SSDF sensor makes the rule more ineffective and tends to make a wrong decision.

6.4.1 SSDF ATTACK: REFERENCE AND MATHEMATICAL EXPRESSIONS

Multiple attackers will collaborate to falsify the data. Each attacker has a unique ID. This attack pattern is a popular collusive attack, especially found in CSS to create errors in the sensing data. This is called as SSDF attack. The SU's sensing function for each case is not permitted due to the shadowing which is shown in Figure 6.1.

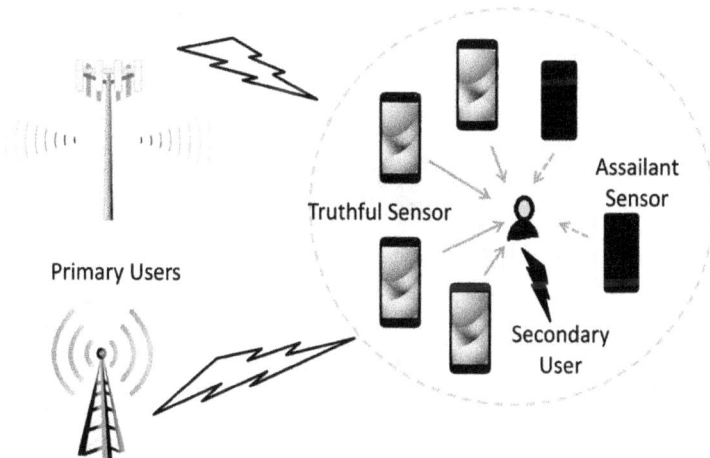

FIGURE 6.1 SSDF sensor at the time of SU data fusion.

This chapter addresses the ill-will SUs of Byzantine/SSDF attacks, classifies them into three types and incorporates with the blind scenario.

The first type is a *'smart ill-will SU'*. That means, if the node senses 1 (occupied) from the primary base station (BS), it will send 0 (vacancy) to the data fusion center (DFC) as a result and contrariwise. The second type is *'always occupied ill-will SU'*. This always sends 1 to the DFC as being an occupied channel. This is not as smart as the first assailant, but it causes a denial of service (DOS) attack, which means that the channel is constantly not available for the truthful SUs. And the third type is *'always vacant ill-will SU'*. This sends 0 to the DFC, which denotes that the SU assumes that the channel is always available. But sometimes the channel is occupied and a collision takes place. The SU uses more energy and time for searching another frame. A truthful node moves its own data and sends its outcome to another node or FC. A Byzantine sensor alters the transmission with some decisions. A mathematical model for the Byzantine attack is assumed as: $P_{j,1}^{T}$, $P_{j,0}^{T}$, $P_{j,1}^{AL}$, $P_{j,0}^{AL}$ $j \in \{0,1\}$

For Truthful nodes:

$$P_{1,0}^{T} = 1 - P_{0,1}^{T} = P^{T}(\hat{S} = 0 | \hat{G} = 1) = 1 \qquad (6.21)$$

$$P_{1,0}^{T} = 1 - P_{0,0}^{T} = P^{T}(\hat{S} = 1 | \hat{G} = 0) = 0 \qquad (6.22)$$

For Byzantine sensors:

$$P_{1,0}^{AL} = 1 - P_{0,1}^{AL} = P^{AL}(\hat{S} = 0 | \hat{G} = 1) = \alpha \qquad (6.23)$$

$$P_{1,0}^{AL} = 1 - P_{0,0}^{AL} = P^{AL}(\hat{S} = 1 | \hat{G} = 0) = \beta \qquad (6.24)$$

Where $P^{AL}(\hat{S} = i | \hat{G} = j)$ is the assailant node probability that sends i as a sensor result and receives j from the global decisions. The actual decision is j. If a node is an attacker, but its ascendants are not, it requires a Byzantine outcome due to another Byzantine which is a neutralized one. The node to the FC will have at least one SSDF or Byzantine, $\sum_{k=1}^{K} \alpha_k \leq 1$. But the number of Byzantines in any other way to the FC cannot be greater than 1. The SSDF CR identity is unknown and is found as follows: α is the probability of a single received detection at the FC which is from a Byzantine. The FC binary hypothesis test is formulated as:

H_0: PU is not active (absent)

$$P_{fa} = P(\dot{R} = j) = (1 - \beta)\rho\left[\left(P_{f}^{T}\right)P_{(1,1)j}^{T} + 1 - \left(P_{f}^{T}\right)P_{(1,0)j}^{T}\right]$$
$$+ \left[\left(P_{f}^{AL}\right)P_{(1,1)j}^{AL} + 1 - \left(P_{f}^{AL}\right)P_{(1,0)j}^{AL}\right]\alpha \qquad (6.25)$$

H_0: PU is active (present)

$$P_{ma} = P\left(\dot{R} = j\right) = (1-\alpha)\rho\left[\left(P_d^T\right)P_{(1,1)j}^T + 1 - \left(P_d^T\right)P_{(1,0)j}^T\right]$$
$$+ \left[\left(P_d^{AL}\right)P_{(1,1)j}^{AL} + 1 - \left(P_d^{AL}\right)P_{(1,0)j}^{AL}\right]\beta \qquad (6.26)$$

$$P_{da} = 1 - P_{ma}$$

FC is mindful of the presence of Byzantine attacks, but it will not differentiate truthful and assailant nodes. An assailant causes the FC to make an incorrect decision regarding the presence or absence of the PU. In the given hypothesis, each SU's sensing result is added to the conditional independent and identical distribution. Therefore, the blind condition [25] can be expressed as:

$$P_f + \rho\left((1 - P_f)\alpha - P_f\beta\right) = 1 - P_m + \alpha(P_m\alpha - (1 - P_m)\beta) \qquad (6.27)$$

So it becomes, $P_{fa} = P_{da}$

By algebraic expression it can be simplified as [25]:

$$\rho = \frac{1}{\alpha + \beta} \qquad (6.28)$$

Nearly a half part of the assailant user can completely blind the FC, when $\alpha = \beta = 1$. It is the minimum rate that makes the FC blind. This assumption shows that the low malicious rate has a always-attack strategy that enables the existing studies [20, 24] to escape the blind problem.

6.5   PROPOSED SYSTEM MODEL

6.5.1   A NOVEL SCHEMA

In 5G assumptions, user devices travel in a large number of wireless devices and they do not know about the honest and ill-will users. The 'novel' method along with the 'improved-apriori' method will solve this problem. Each frame of cognition contains two stages: the sensing and the accessing phase of the spectrum. The accessing phase relies on the universal report results. The SU will access if the universal decision has the result 'the PU is absent'. Otherwise, the SU has to wait for the upcoming frame for the process to be done. This is a general cognitive process. If the SU approaches the PU channel, one of two tasks will occur to transfer the information successfully. Otherwise, it may collide with the PU communication. If the PU is not available, the universal report is right, and 'success' is possible; or else, it may cause failure. This can be used as a feedback for the defense performance.

If the SU waits for the upcoming frame, one of two tasks will occur: i.e., an 'idle' or 'busy' PU channel. It may lose due to a fake universal decision report. The restricted SU sensing execution will be upgraded by expanding the time of sensing.

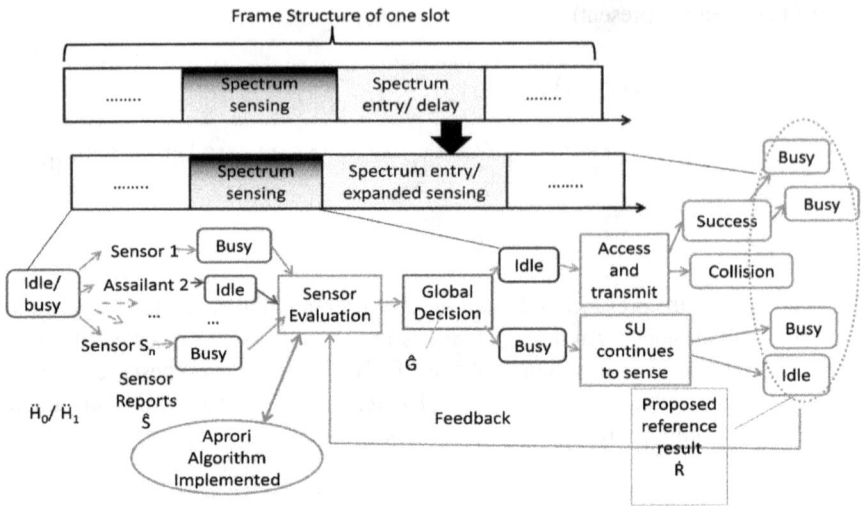

FIGURE 6.2 An execution of an 'improved-apriori' algorithm along with the 'novel' method, where $\ddot{H}_0 = \ddot{H}_1$ indicates the original channel condition, $\acute{R}$ denotes the proposed result, and $\hat{G}$ is the universal decision report.

But, at the time of the waiting stage, the SU tracks the sensing process and enlarges the restricted SU's sensing results as shown in Figure 6.2. The enlarged sensing outcome is used as usable feedback.

6.5.2 PROPOSED METHOD

6.5.2.1 An "Improved-apriori" Algorithm

This chapter uses the "improved-apriori" (i-apriori) algorithm for the possible improved detections which are higher than the existing ones. The K-out-of-N rule is incorporated in the pseudocode below, as follows:

```
Apriori (A, ∈)
(SU)₁ = find_frequent_1_itemsets(T);
// construct N₂by self - join
N₂ = (SU)₁∞(SU)₁
(SU)₂ = items in N₂ ≥ min_sup;
For (k = 3;(SU)_{f-1} ≠ ∅; k + +)
{// further prune (SU)_{f-1}
Prune1((SU)_{f-1});
(SU)_x ∈ (SU)_f,(SU)_x ∈ (SU)_f;
If((SU)_i [1] = (SU)_j [1] ∧ (SU)_i [2] = (SU)_j [2]
∧...(SU)_i [k - 2] = (SU)_j [k - 2] ∧ (SU)_i [k - 1]
< (SU)_j [k - 1])
N = (SU)_i ∪ (SU)_j ;
```

```
If (k - 1) - subsets s of n ∉ (SU)_{f-1}
then delete N from N_k;
N_k = n ∪ N_k;
// Intersection between the target transaction
TID set to calculate the support
(SU)_f k = New - quick_support_count
(N_k, k counter / (SU)_u_S_f);
Answer = U_k(SU)_k;
New_quick_support_count(N_k, k counter / (SU)_u_S_f);
{for all S_f N ∈ N_f
N.(SU)_u _ S_f = N_{k-1}.(SU)_u _ S_f n N_1.(SU)_u _ S_f
N.sup = Length(N_k.(SU)_u_S_f);
If N.sup < min_sup
Delete N from N_k
(SU)_f k = {N ∈ N_k |N.sup ≥ min_sup}}
Prunel (SU)_f
for all S_f (SU)_f ∈ (SU)_k;
If count (SU)_f in (SU)_k ≤ k;
Then delete all (SU)_j from (SU)_k
return (SU)_f
```

In this code, all the SU sensor reports $\hat{S}$ are divided into i and j parts. Each part takes a maximum of two reports in $\hat{S}$. Let k be the counter that takes these two nodes to check the similarity. If the nodes are similar, it removes the malicious node. Let S_n be the number of sensor and N_k counts the number of SUs, which are not similarly under checking in k counter. Let f be the parameter that selects any of these two nodes and S_f is the sensor that takes the selected two nodes. Then, N_f is the counter that counts the f pair series and $(Su)_f$ denotes the entire SU pair series. Let u be the parameter that takes an apriori algorithm (A, ε) for parameter N_k and $(SU)_f$. Then it is counted as $(SU)_u$ and $(SU)_k$ is the number of SUs k counter.

To annul the repeated scan of distributed data, the improved-apriori (also known as i-apriori) algorithm is suggested. The procedure is as follows [26]:

1. The scanned reports are considered, to get the k counter for a SU.
2. Rationalize $(SU)_f-1$, before the SU report N_k comes. Then, count all the SU occurred in N_f-1, and delete SUs, which is less than $k-1$ [15].
3. Through the convergence strategy, count the nodes of right SU using N_k based on the k counter of $(SU)_{f-1}$ and SU of N_l.
4. Then cease the algorithm, if $|(SU)_f| \leq k$ [16].

This algorithm reduces the number of assailant SUs and the time to count all the truthful nodes decreases. The i-apriori contains a receiver id (k counter) for the distributed CR systems [26]. In the algorithm, T denotes the item set and *min_sup* is the parameter which helps to know about the minimum support for honest SU. In the i-apriori algorithm, the parameter N takes the $(SU)_i$ and $(SU)_j$. The infinite number

TABLE 6.2

The Receiving k Counter of i-apriori

k counter/$(SU)_u$	S_f	k counter/$(SU)_u$	Secondary Users S_f
$K+1$	SU_1, SU_2	$K+6$	SU_1, SU_3, SU_5
$K+2$	SU_2, SU_3	$K+7$	SU_2, SU_3
$K+3$	SU_1, SU_2	$K+8$	SU_3, SU_4
$K+4$	SU_1, SU_2, SU_3, SU_4	$K+9$	SU_4, SU_6
$K+5$	SU_1, SU_2, SU_3	$K+10$	SU_3, SU_5

of SUs will be united up to N_2 and R is the outcome to get the right SU in the whole SU pairs.

A 10 k counter and a 6 SU set are shown in Table 6.2. The minimum level of SU taken is 2.

6.5.3 IMPLEMENTATION OF THE PROPOSED METHOD

The $\dot{R}$ helps the SU to re-assess the restricted sensing execution for every individual sensor. The false alarm probability of the proposed one is calculated as:

$$P_f^{Pr} \triangleq P_0\left(\dot{R}=1\middle|\ddot{H}_0\right) = P_0\left(\dot{R}=1\middle|\ddot{H}_0, \hat{G}=0\right)$$

$$. P_0\left(\hat{G}=0\middle|\ddot{H}_0\right) + P_0\left(\dot{R}=1\middle|\hat{G}=0, \ddot{H}_0\right). P_0\left(\hat{G}=0\middle|\ddot{H}_0\right)P_{fa} \tag{6.29}$$

$$= 0 + P_f^{Ex}. Q_f P_{fa} \tag{6.30}$$

The universal report of $\hat{G} \, \varepsilon \, \{0(\text{inert}), 1(\text{full})\}$ and the probability of false alarm is given by, $P_f^{Pr} \triangleq P_0\left(\hat{G}=1, \ddot{H}_0\right)$. Then $\dot{R} \, \varepsilon \, \{0(\text{inert}), 1(\text{full})\}$ is the result of the proposed i-apriori method, and $P_{md}^{Pr} \triangleq P_0\left(\hat{G}=0, \ddot{H}_1\right)$ is a missed-detection probability. It contains $P_0\left(\dot{R}=\ddot{H}_i \middle| \hat{G}=0\right) = 1$. Otherwise, the SU executes an enlarged sensing of the spectrum, even when the decision is $\hat{G}=1$. The result of sense is over at the total slit, then $P_0\left(\dot{R}=1\middle| \hat{G}=1, \ddot{H}_0\right) = P_f^{Ex}$. Likewise, the suggested miss-detection probability is:

$$P_{md}^{Pr} \triangleq P_0\left(\ddot{H}_1\middle|\dot{R}=0\right) = P_0\left(\hat{G}=0\middle| \ddot{H}_1\right). P_0$$

$$\left(\ddot{H}_1\middle|\dot{R}=0, \hat{G}=0\right) + P_0\left(\dot{R}=0\middle|\hat{G}=1, \ddot{H}_1\right)$$

$$. P_0\left(\hat{G}=1\middle| \ddot{H}_1\right)P_{ma} \tag{6.31}$$

$$= 0 + P_{md}^{Ex}. \left(1-Q_d\right)P_{ma} \tag{6.32}$$

6.5.3.1 Performance Evaluation of the Accuracy

The adjacent existing sources ($\ddot{H}_1$ (busy) or else $\ddot{H}_0$ (idle)) are evaluated in the spectrum sensing execution, ie., in the probability of miss-detection and false alarm. The preferred one is the probability of error [21]. The probability of error P_e in (6.32) is given as:

$$P_e = P_0\left(\ddot{H}_0\right)P_G^f \cdot P_f^{Ex}P_{at} + P_0\left(\ddot{H}_1\right)\left(1 - P_G^m\right) \cdot P_{md}^{Ex}P_{ta} \tag{6.33}$$

$$P_e = P_0\left(\ddot{H}_0\right)P_G^f \cdot P_0\left(\frac{\varphi - \partial_0}{\delta_1}\right) \tag{6.34}$$

The probability that the real channel condition is P_0 ($\ddot{H}_0$) and P_0 ($\ddot{H}_1$). Due to the enlarged sensing of the spectrum, ∂_0, ∂_1 and δ_0, δ_1 are r times of ∂_0, ∂_1 and δ_0, δ_1 are applied in Equation (6.17) respectively. The values are obtained from the i-apriori learning strategy. The proportion for the total sensing slot with the real sensing slot is r. To find the detection performance, the threshold is φ. Then, it contains the following suggestion:

Suggestion 1: The threshold φ_p is the defense scheme of SU that will reduce the error probability P_e as follows:

$$\varphi_p = \min P_e(\varphi) \tag{6.35}$$

$$= \begin{cases} x_l, \; k \leq 0 \\ \arg\min P_e\;(\varphi), \; j^2 - 4ik > 0, \; k > 0 \\ \varphi \in \{0, x_l\} \\ 0, \; j^2 - 4ik < 0 \end{cases} \tag{6.36}$$

And,

$$x_l = \frac{-j + \sqrt{j^2 - 4ik}}{2i}, \; i = \frac{1}{\delta_0^2} - \frac{1}{\delta_1^2}, \; j = 2\left(\frac{\partial_1}{\delta_1^2} - \frac{\partial_0}{\delta_0^2}\right) \tag{6.37}$$

and

$$k = \frac{\partial_0^2}{\delta_0^2} + \frac{\partial_1^2}{\delta_1^2} \; 2\ln\left(\frac{P\left(\ddot{H}_1\right)\left(1 - P_G^m\right)\delta_0}{P\left(\ddot{H}_0\right)\left(P_G^f\right)\delta_1}\right) \tag{6.38}$$

The above formulation of k is estimated by the i-apriori algorithm. Then the estimation problem is solved by the maximum likelihood estimation, which follows:

$$max\; L(\acute{P}_f^x, \acute{P}_d^x) = max \prod_{j=1}^{S_t} p\left(y_i(j); P_f^{Ex}P_d^{Ex}\right) \tag{6.39}$$

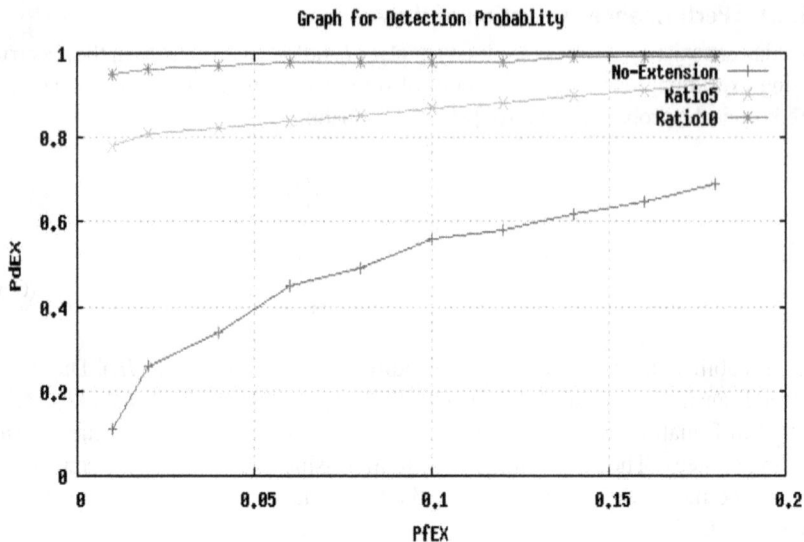

FIGURE 6.3 Enlarged sensing indicates the probability of detection P_{md}^{Ex} and probability of false alarm P_f^{Ex} for the restricted SNR is −10 dB.

For a proof, see Appendix I in this chapter, the real universal decision (UD), i.e., P_f^G and P_{md}^G. Then $\varphi = 0$ for the suggested source, $P_f^{Ex} = 1$, $P_{md}^{Ex} = 1$. Figure 6.3 indicates the suggested sources under different assailant probability P_m and an assailant populations in k counter.

Let Φ be the source for probability of false alarm ε_f and the probability of miss-detection ε_{md}. The assigned sensor is $x_l \, \varepsilon \, \{T, AL\}$. The sensor outcome assigns x_l for $\hat{S}^{xl}$ whose identification operation is calculated as (P_{md}^{xl}, P_f^{xl}). The possibility for $\hat{S}^{xl} \neq \Phi$ is calculated as:

$$\varepsilon^{xl} = P_0(\hat{S}^{xl} \neq \Phi) \tag{6.40}$$

$$= P_0\left(\ddot{H}_1\right) P_0\left(\hat{S}^{xl} \neq \Phi \mid \ddot{H}_1\right) + P_0\left(\ddot{H}_0\right) P_0\left(\hat{S}^{xl} \neq \Phi \mid \ddot{H}_{s0}\right) \tag{6.41}$$

$$= P_0\left(\ddot{H}_1\right) \left[P_{md}^{xl} + \left(1 - 2P_{md}^{xl}\right) \varepsilon_{md} \right] + P_0\left(\ddot{H}_0\right) \left[P_f^{xl} + \left(1 - 2P_f^{xl}\right) \varepsilon_f \right] \tag{6.42}$$

In (6.42), ε^{xl} denotes the actual execution of the sensor and is assigned as x_l. The universal report [9], t_0th for the sensor reputation value ω^{xl} is calculated as:

$$\omega^{xl}(t_0) = \omega^{xl}(t_0 - 1) + f \tag{6.43}$$

If $f = 0$ then $\hat{S}^{x1}(t) \neq \Phi(t)$ or $f = 1$ for $\hat{S}^{x1}(t) = \Phi(t)$. The binomial distribution is modified and a Gaussian estimation [9] is given as:

$$\omega^T \sim S(\partial_{AL}, \delta^2_{AL}) \tag{6.44}$$

$$\omega^{AL} \sim S(\partial_T, \delta^2_T) \tag{6.45}$$

Where, $\partial_{AL} = S_t \times^{AL}, \delta^2_{AL} = S_t \times^{AL}(1 - \times^{AL}), \times^{AL} = S_t \times^T, \delta^2_T = S_t \times^T(1 - \times^T)$
The assigned sensor is:

$$\omega^{x1} \begin{array}{c} x_1 = AL \\ >< \quad \theta \\ x_1 = T \end{array} \tag{6.46}$$

The threshold is θ and P_{at} is an assailant sensor probability that is incorrectly assigned as truthful.

Let P_{ta} be the truthful sensor probability that is incorrectly assigned as an assailant one. Both of them are separately calculated as:

$$P_{at} \triangleq P_0\left(\omega^T \geq \theta\right) = \int_{\theta}^{+\infty} f\left(\frac{x_1 - \partial_T}{\delta^2_T}\right) dx_1 \tag{6.47}$$

$$P_{ta} \triangleq P_0\left(\omega^{AL} < \theta\right) = \int_{-\infty}^{\theta} f\left(\frac{x_1 - \partial_{AL}}{\delta^2_{AL}}\right) dx_1 \tag{6.48}$$

Let π^{AL} and π^T indicate the rate of truthful and assailant sensors. Then, $\pi^{AL} + \pi^T = 1$ that contains the possibility of wrongly detecting sensors which is assigned as P_{error} as follows:

$$P_{error} = \pi^{AL} P_{at}(\theta) + \pi^T P_{ta}(\theta) \tag{6.49}$$

$$= P_{error}\left(W_m + \Delta_m, W_f + \Delta_f\right) - P_{error}(W_m, W_f) \geq 0 \tag{6.50}$$

$$\forall \Delta_m, \Delta_f \geq 0$$

A maximum posteriori probability rule simplifies the K-out-of-N rule. Therefore, the global probability of false alarm and miss-detection [25] under the rule for the data fusion is given as:

$$Q_f = \sum_{i=K}^{N} \binom{i}{N} P_{at}^n (1 - P_{at})^{N-n} \tag{6.51}$$

$$Q_m = \sum_{i=K}^{N} \binom{i}{N} P_{ta}^n (1 - P_{ta})^{N-n} \tag{6.52}$$

6.6 DISCOURSE AND PERFORMANCE RATING

6.6.1 BASIC SIMULATION CONFIGURATION

The local SNR for the SUs is taken as -10dB. The chance for the PU channel equaling the full origin is 0.1. The truthful detector performance is adjusted as $P_f^T = 0.1$ and $P_d^T = 0.9$. The probability of the assailant is P_m and an assailant detector ranges from 0–1. The time bandwidth for the extended sensing of SUs ranges from 100 to 500 bps. The probability of false alarm is 0.08. Then $P_e = 0.01$, constitutes the worse for WiMax network [9] and $P_e = 0.001$ is adopted for Wi-Fi networks that do not include the results of $P_e = 0.1$, which indicates the rough wireless conditions.

Figures 6.4 and 6.5 represent the enhanced performance of the proposed method compared to different existing techniques with different attack probabilities and populations. The proposed method greatly reduces the error probability of a CSS.

From Figures 6.4 and 6.5, we can conclude that the sequential 0/1 scheme is the competing technique to be compared with the proposed method. Therefore, for all the following simulations, s0/1 strategy is implemented and compared with the i-apriori technique.

Figures 6.6 and 6.7 represent the performance of the s0/1 technique and the i-apriori method given enlarged sensing. The proposed technique performs better than the existing s0/1.

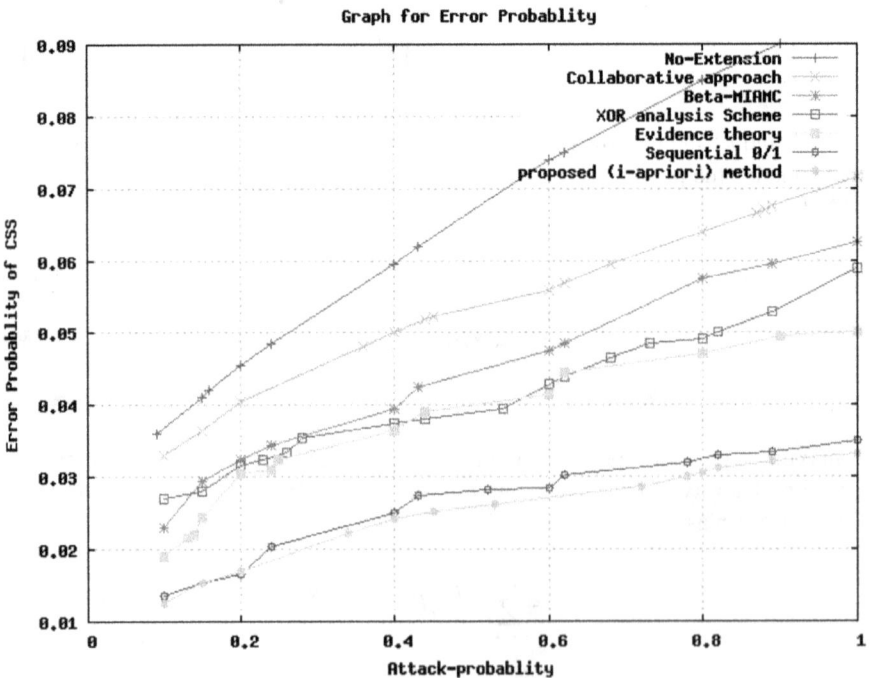

FIGURE 6.4 Performance of i-apriori with the existing methods in different attack probabilities.

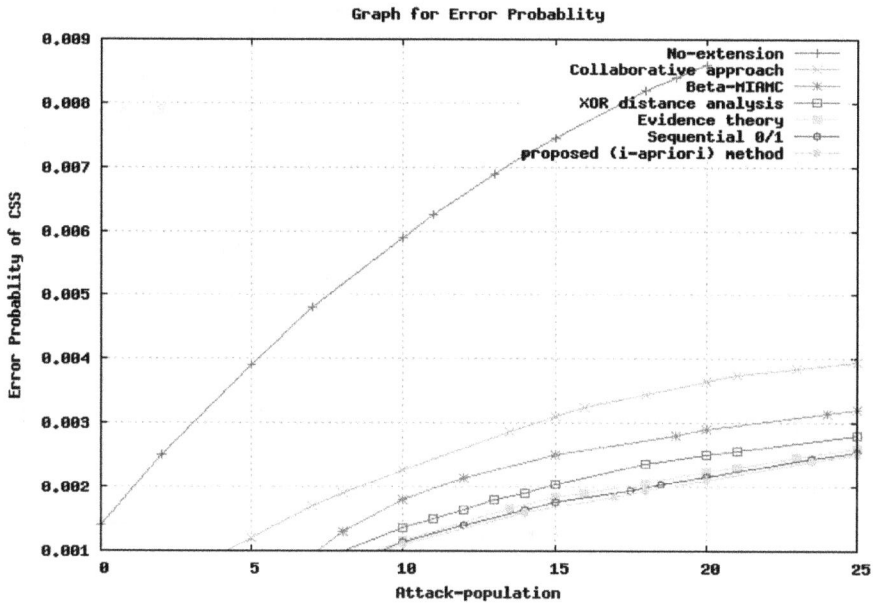

FIGURE 6.5 Performance of i-apriori with the existing methods in different attack populations.

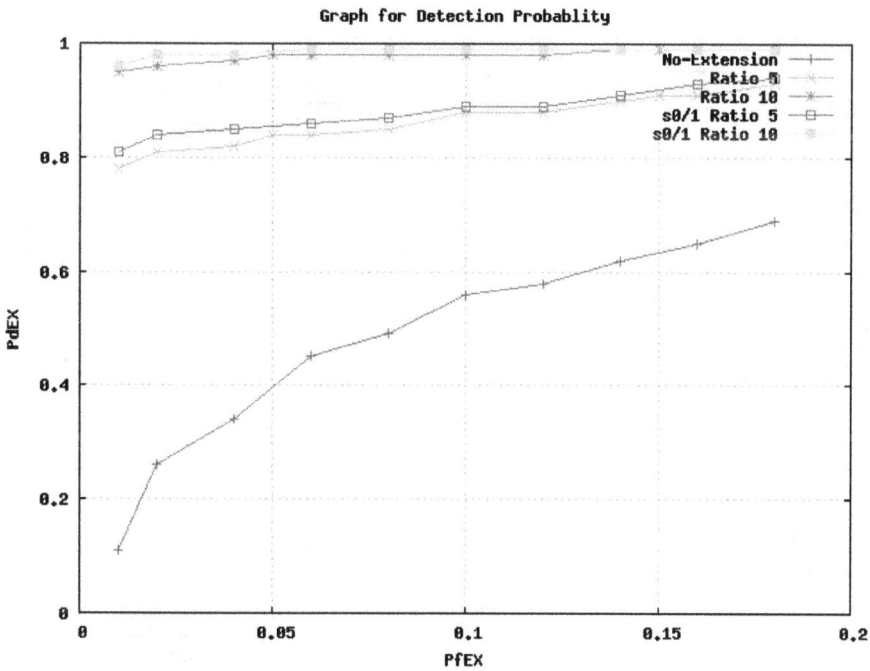

FIGURE 6.6 Performance of existing sequential 0/1 method in enlarged sensing detection.

Graph for Detection Probablity

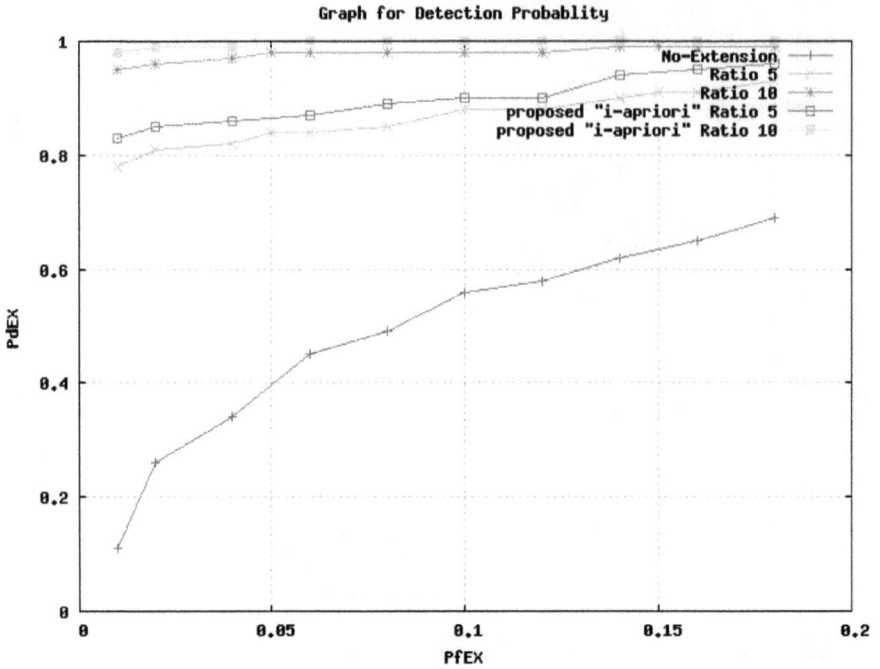

FIGURE 6.7   Performance of proposed i-apriori method in enlarged sensing detection.

Transmission rate over time

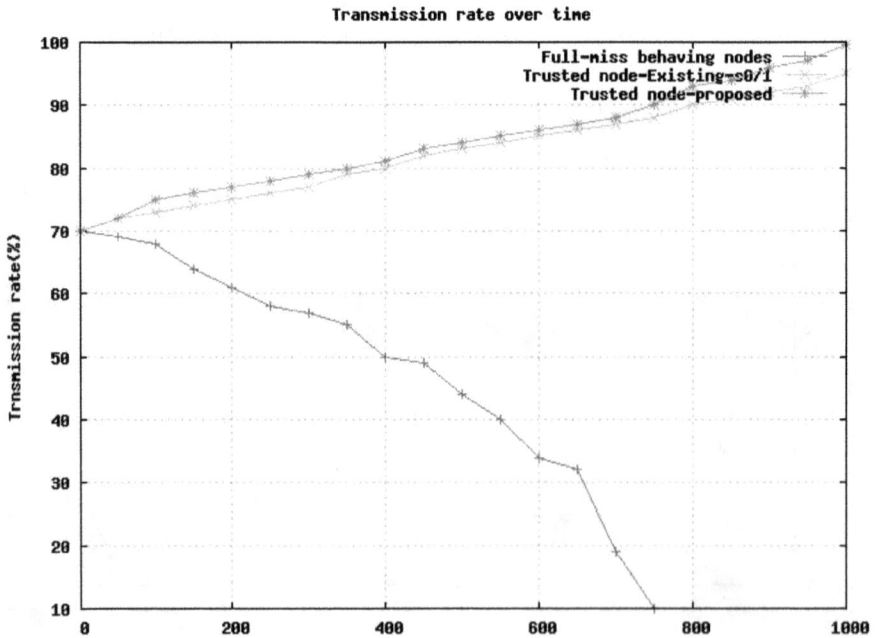

FIGURE 6.8   Performance of transmission rate over time.

Correct sensing ratio

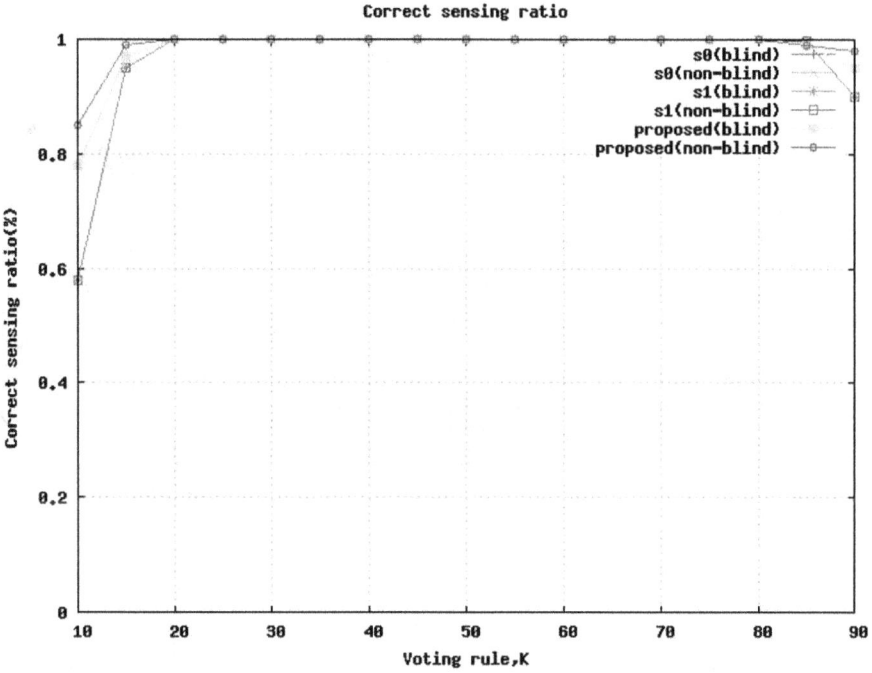

FIGURE 6.9 Correct sensing ratio of existing s0/1 and proposed i-apriori scheme in blind and non-blind scenario.

In Figure 6.8, the trusted node transmission rate is calculated from the existing method of $\bar{N}_{s0/1}$ in Equations (6.5 and 6.6) and from the proposed method Equation (6.37) and (6.38) through the k counter.

In Figure 6.9, for voting rule k, the existing and proposed result will be taken according to the blind and non-blind scenario. The s0/1 result is taken from article [21] and the proposed i-apriori scheme is implemented according to the blind scenario in Equation (6.27). For a non-blind scenario, only α and β values are implemented by eliminating the ρ value in (6.25) and (6.26).

6.7 CONCLUSIONS

This chapter focuses on two new methods for the security of CSS in CRN, which results in the security of 5G-CR technologies. The proposed method confronts SSDF attack in CR as well as future trends of modern technologies like IoT, 5G, Vanet, etc. The chapter's simulation is done in NS-3, a numerical simulation that confirms that the proposed method has better performance, especially in the case of a major or minor assailant sensor.

REFERENCES

1. J. Rodriguez, *"Fundamentals of 5G Mobile Networks: Security for 5G Communications,"* West Sussex, UK: John Wiley & Sons Ltd., 2015, pp. 207–219.
2. E. Hossain, and M. Hasan, "5G Cellular: Key Enabling Technologies and Research Challenges," *IEEE Instrumentation & Measurement Magazine*, vol. 18, no. 3, June 2015, pp. 11–21.
3. J. Mitola, *"Cognitive Radio: an Integrated Agent Architecture for Soft-Ware Define Radio,"* Ph.D. dissertation, Dept. Tele-info., Royal Inst. Of Tech. (KTH), Stockholm, Sweden, 2000.
4. A. Mesodiakaki, F. Adelantado, L. Alonso, and C. Verikoukis, "Energy Efficiency Analysis of Secondary Networks in Cognitive Radio Systems," *IEEE Int. Conf. on Comm., Selected Areas in Communications Symposium*, Budapest, Hungary, June 9–13, 2013, pp. 4115–4119.
5. H. Li, X. Cheng, K. Li, C. Hu, and N. Zhang, "Robust Collaborative Spectrum Sensing Schemes in Cognitive Radio Networks," *IEEE Transactions on Parallel and Distributed System*, vol. 25, no. 8, August 2014, pp. 2190–2200.
6. E. Nurellari, D. McLernon, and M. Ghogho, "A Secure Optimum Distributed Detection Scheme in Under-Attack Wireless Sensor Networks," *IEEE Transaction on Signal and Information Processing Over Networks*, vol. 4, no. 2, June 2018, pp. 325–337.
7. A.H.S. Magdalene, and L. Thulasimani, "Analysis of Spectrum Sensing Data Falsification (SSDF) Attack in Cognitive Radio Networks: A Survey," *Journal of Science and Engineering Education*, vol. 2, 2017, pp. 89–100.
8. H. Luan, O. Li, and X. Zhang, "Cooperative Spectrum Sensing with Energy Efficient Sequential Decision Fusion Rule," *IEEE Wireless and Optical Communication Conference*, pp. 1–4, 2014.
9. A.A. Alkheir, and H.T. Mouftah, "Sequential Hard-Decision Fusion for Agile Cooperative Spectrum Sensing," *IEEE International Conference on Communication Workshop*, pp. 1014–1019, 2015.
10. S. Peng, W. Zheng, R. Gao, and K. Lei, "Fast Cooperative Energy Detection Under Accuracy Constraints in Cognitive Radio Networks," *Wireless Communications & Mobile Computing*, pp. 1–8, 2017.
11. C.I. Badoi, N. Prasad, V. Croitoru, and R. Prasad, "5G Based on Cognitive Radio," *Wireless Personal Communication Journal*, vol. 57, no. 3, April 2011, pp. 441–464.
12. Ericsson. *5G Radio Access*. Uen 284 23-3204 Rev C. Stockholm, 2016.
13. R. Chen, J.M. Park, Y.T. Hou, and J.H. Reed, "Toward Secure Distributed Spectrum Sensing in Cognitive Radio Networks," *IEEE Communication Magazine*, vol. 46, no. 4, April 2008, pp. 50–55.
14. V. Sucasas, A. Radwan, S. Mumtaz, and J. Rodriguez, "Effect of Noisy Channels in MAC-Based SSDF Counter-Mechanisms for 5G Cognitive Radio Networks," *International Symposium on Wireless Communication System*, Brussels, Belgium, August 25–28, 2015, pp. 1–5.
15. J. Lu, P. Wei, and Z. Chen, "A Scheme to Counter SSDF Attacks Based on Hard Decision in Cognitive Radio Networks," *WSEAS Transaction on Communication*, vol. 13, 2014, pp. 242–248.
16. S. Kim, H. Cha, J. Kim, S.W. Ko, and S.L. Kim, "Sense-and-Predict: Harnessing Spatial Interference Correlation for Cognitive Radio Networks," *IEEE Transactions on Wireless Communications*, vol. 99 (accepted), April 2019, pp. 1–17.
17. F. Song, Y.T. Zhou, L. Chang, and H.K. Zhang, "Modeling Space-Terrestrial Integrated Networks with Smart Collaborative Theory," *IEEE Networks*, vol. 33, no. 1, January 2019, pp. 51–57.

18. S. Bhattacharjee, R. Keitangnao, and N. Marchang, "Association Rule Mining for Detection of Colluding SSDF Attack in Cognitive Radio Networks," International Conference on Computer Comm. & Info., Coimbatore, India, 2016.
19. R. Amutha Priya, and S. Nandhakumar, "Attack Prevention for Spectrum Sensing Data Falsification Attacks in Cognitive Radio Networks Using Arc," *International Journal of Advanced Research in Science, Engineering and Technology*, vol. 2, no. 3, March 2015, pp. 486–490.
20. M. Khasawneh, and A. Agarwal, "A Collaborative Approach Towards Securing Spectrum Sensing in Cognitive Radio Networks," *Procedia Computer Science*, vol. 94, no. 2016, December 2016, pp. 302–309.
21. J. Wu, Y. Yu, T. Song, and J. Hu, "Sequential 0/1 for Cooperative Spectrum Sensing in the Presence of Strategic Byzantine Attack," *IEEE Wireless Communications Letters*, vol. 8, no. 2, April 2019, pp. 500–503.
22. J. Feng, M. Zhang, Y. Xiao, and H. Yue, "Securing Cooperative Spectrum Sensing Against Collusive SSDF Attack Using XOR Distance Analysis in Cognitive Radio Networks", *Sensors*, vol. 18, no. 2, January 2018, pp. 1–14.
23. J. Zhang, L. Cai, and S. Zhang, "Malicious Cognitive User Identification Algorithm in Centralized Spectrum Sensing System", *Future Internet: MDPI Journal*, vol. 9, no. 79, November 2017, pp. 1–13.
24. H. Wang, Y. Li, and T.C. Chang, "An Enhanced Cooperative Spectrum Sensing Scheme for Anti-SSDF Attack Based on Evidence Theory," *Microsystem Technologies*, vol. 24, no. 6, June 2018, pp. 2803–2811.
25. J. Wu, T. Song, Y. Yu, C. Wang, and J. Hu, "Generalized Byzantine Attack and Defense in Cooperative Spectrum Sensing for Cognitive Radio Networks" *IEEE Access*, vol. 6, August 2018, pp. 53272–53286.
26. X. Yuan, "An Improved A Priori Algorithm for Mining Association Rules," *AIP Conference Proceeding Journal*, vol. 1820, no. 1, Mar. 2017, pp. 080005-1–080005-6.

APPENDIX I: PROOF OF ERROR PROBABILITY, P_E

From (6.12), taking the differential with respect to φ, we have:

$$\frac{\partial P_e}{\partial \varphi} = P_0\left(\ddot{H}_0\right) P_G^f \frac{\partial P_f^{Ex}}{\partial \varphi} + P_0\left(\ddot{H}_1\right)\left(1 - P_G^m\right) \frac{\partial P_{md}^{Ex}}{\partial \varphi}$$

Let (6.14) be 0, we have:

$$2\ln\left(\frac{P_0\left(\ddot{H}_1\right)\left(1 - P_G^m\right)\delta_0}{P_0\left(\ddot{H}_0\right)P_G^f\delta_1}\right) = \frac{1}{\delta_1^{\,2}}\left(\varphi - \partial_1\right)^2 - \frac{1}{\delta_0^{\,2}}\left(\varphi - \partial_0\right)^2$$

Here,

$$h\left(\varphi\right) = \frac{1}{\delta_0^{\,2}}\left(\varphi - \partial_0\right)^2 - \frac{1}{\delta_1^{\,2}}\left(\varphi - \partial_1\right)^2 + 2\ln\left(\frac{P_0\left(\ddot{H}_1\right)\left(1 - P_G^m\right)\delta_0}{P_0\left(\ddot{H}_0\right)P_G^f\delta_1}\right)$$

$$\left(\frac{1}{\delta_0{}^2} - \frac{1}{\delta_1{}^2}\right)\varphi^2 + 2\left(\frac{\partial_1}{\delta_1{}^2} - \frac{\partial_0}{\delta_0{}^2}\right)\varphi + \frac{\partial_1{}^2}{\delta_1{}^2} - \frac{\partial_0{}^2}{\delta_0{}^2} + 2\ln\left(\frac{P_0(\ddot{H}_1)(1-P_G^m)\delta_0}{P_0(\ddot{H}_0)P_G^f\delta_1}\right)$$

$$\triangleq i\varphi^2 + j\varphi = k$$

Where

$$i = \frac{1}{\delta_0^2} - \frac{1}{\delta_1^2}, \; j = 2\left(\frac{\partial_1}{\delta_1^2} - \frac{\partial_0}{\delta_0^2}\right) \text{ and } k = \frac{\partial_0^2}{\delta_0^2} -$$

$$\frac{\partial_1^2}{\delta_1^2} + 2\ln\left(\frac{P(\ddot{H}_1)(1-P_G^m)\delta_0}{P(\ddot{H}_0)(P_G^f)\delta_1}\right)$$

Since, $\delta_0 > \delta_1$, $\therefore i > 0$

Similarly, at the same time:

$$\frac{\partial_1}{\delta_1^2} = \frac{\alpha+1}{2(2\beta+1)}, \frac{\partial_0}{\delta_0^2} = \frac{1}{2} \; \therefore b < 0$$

Case 1: $k \leq 0$, then have, $j^2 - 4ik > 0$, hence the highest outcome is the optimum value φ_p.

Case 2: $k > 0$, then if, $j^2 - 4ik \geq 0$, φ_p is one of zero and a larger solution to $h(\varphi)$, which makes $P_e(\varphi)$ get a smaller value, if $j^2 - 4ik < 0$, then $\varphi_p = 0$.

7 Detection of Retinopathy of Prematurity Using Convolution Neural Network

Deepa Dhanaskodi and Poongodi Chenniappan
Bannari Amman Institute of Technology

CONTENTS

7.1 INTRODUCTION

7.1.1 RETINOPATHY OF PREMATURITY

Retinopathy of prematurity (ROP) is an eye disease observed in babies delivered prematurely; it is considered to be a serious issue that must be treated because it causes childhood blindness. ROP is also known as retrolental fibroplasias (RLF). Initially, it was believed that oxygen therapy caused the condition, since it occurs in newborn, especially in premature, babies who received oxygen therapy. The role of supplemental oxygen in ROP has since been disproven. Vascularization is the process that is responsible for the normal development of blood vessels in the retina; it is not completed until a baby reaches full term of 40 weeks in uteri. In premature babies, the retina has not fully developed. After the birth of premature babies, if retinal vascularization completes outside the uterus, the retinal vessels may stop growing, or grow abnormally. ROP will also arise when these vessels develop in an abnormal way (Figures 7.1 and 7.2).

7.1.2 CAUSES OF ROP

- ROP happens when anomalous veins develop and spread all through the retina, the tissue that lines the back of the eye.
- These anomalous veins are delicate and can leak, scarring the retina, and pulling it out of position. This causes a retinal detachment. Retinal detachment is the fundamental driver of visual impedance and visual impairment in ROP.
- The eye begins to develop at around about four months of pregnancy, when the veins of the retina start to shape at the optic nerve in the back of the eye. If this process is affected, it will cause ROP.

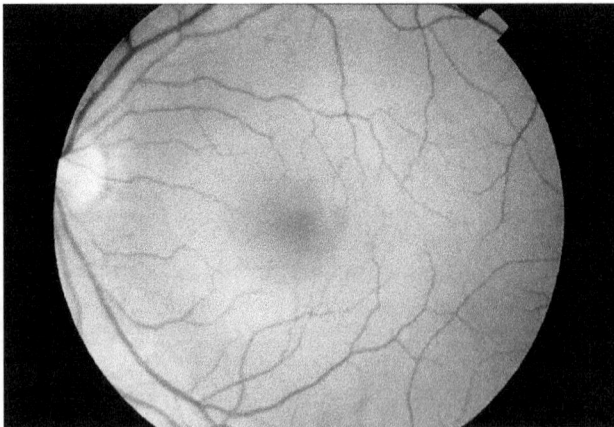

FIGURE 7.1   Normal eye.

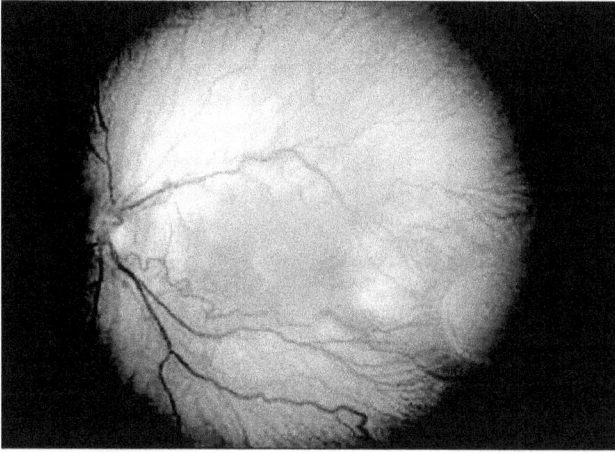

FIGURE 7.2   ROP-affected eye.

- The veins develop steadily toward the edges of the developing retina, providing oxygen and supplements. Insufficient blood flow will cause ROP.
- At the point when a child is brought into the world fullterm, the retinal vein development is generally finished (the retina ordinarily wraps up half a month to a month after birth). In any case, if a child is born prematurely, before these veins have arrived at the edges of the retina, typical vessel development may stop. The edges of the retina the outskirts may not get enough oxygen and supplements.

7.1.3   STAGES OF ROP

Stage I Represents mildly strange vein development. The condition is often associated with transient visual impairment if identified it can be medicated at earlier stage [1].

Stage II Specifies moderately unusual vein development, again often transient, which eventually resolves spontaneously.

Stage III is severely irregular vein development. The unusual veins grow towards the center of the eye instead of the normal growth pattern. There is no treatment for the children who suffer in this stage. The threatened outcome with the diseases in stage III is retinal detachment. Some infants in this stage escape with regular vision.

Stage IV occurs when the retina partially detaches from its original position. Scars are produced by bleeding and abnormal vessels, and they pull the retina away from the wall of the eye.

Stage V results in a fully disconnected retina. The infant is considered to have serious visual disability and even visual deficiency.

7.2   LITERATURE SURVEY

In 2005, Gole [2] introduced the concept of a more virulent form of retinopathy observed in babies that is very aggressive in posterior ROP compared to conventional cases [3–5]. The new classification describes an intermediate level of increased disease between normal posterior pole vessels and the increased disease, and a practical tool for estimating the extent of the identified zone.

In 2009, Wilkinson [6] suggested that with small modification the GoogLeNet can be pertained as a ROP detector.

In 2013, Fleck investigated the role of oxygen in ROP [7]. The authors found a higher risk of severe ROP results due to high levels of oxygen saturation. From their results they recommended that infants who have gestational age less than 28 weeks should maintain more than 90% of oxygen saturation.

7.3   PROPOSED ALGORITHM

7.3.1   CONVOLUTION NEURAL NETWORK

Artificial neural networks (ANN) are used in various classification tasks involving image, audio, speech, etc. Different kinds of neural networks are used for various functions; ANN performs well when machine leaning is involved for image classification. Computer vision with deep learning [8] is becoming more conventional using convolutional neural network (CNN). Image and video recognition, classification, and image analysis tasks can be efficiently carried out using CNN. The details of the layers in the proposed work are discussed below.

7.3.1.1   Input Layer
- The input layer takes the data into the system for classification.
- The number of layers depends on the features that have to be compared for classification.

7.3.1.2   Hidden Layer
- The hidden layer takes the data from input layer. Based on the application and the system model the number of hidden layers can be increased. There are several hidden layers relying upon the model based on information size.
- Every hidden layer will have different numbers of neurons; the number of neurons will be usually larger than the quantity of options (Figure 7.3).
- The output from every layer is computed by matrix operations based on the output of the previous layer, weight value of the current layer, and the bias value followed by an activation function that makes the network nonlinear. The following are the layers present in CNN for efficient classification:
  - CONVOLUTION LAYER:
    - Primarily used to extract features from the input image. The inputs of the convolution layer are the image matrix and filter.

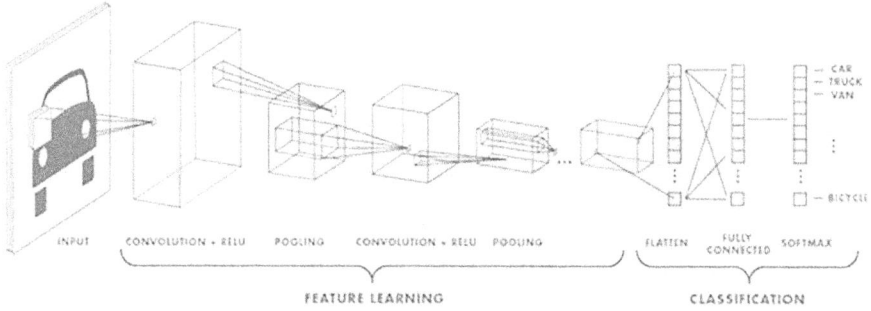

FIGURE 7.3 CNN sequence for classification [9].

- RECTIFIED LINEAR UNIT:
  - The rectified linear unit (ReLU) permits quicker and easier training of a data. It is an activation function. It changes the negative value in image matrix to zero and maintains the positive values.
- POOLING LAYER:
  - When the image size is very large, a pooling layer is used to reduce the number of parameters in the image matrix. The types of pooling layers are max pooling, min pooling, and average pooling.

7.3.1.3 Output Layer

The output of the final hidden layer is given to a logistical function like sigmoid that converts the output of each class into probability of each class.

7.4 PROPOSED WORK FLOW FOR ROP DETECTION

7.4.1 SURVEY AND LOAD IMAGE DATA

- Eye images with and without defects are loaded as dataset for training and testing of ROP. The stored images are then verified during classification process using CNN.
- A picture dataset can store enormous amounts of information, including information which is not present in the memory, and this will productively group the pictures during the preparation of the convolution system.

7.4.2 SPECIFY TRAINING AND VALIDATION SETS

- This step partitions the image information into preparation and evaluation datasets, with the goal that every classification in the preparation dataset includes numerous pictures, and the approval dataset includes the rest of the pictures.
- Each label is divided and stored in the data store for preparing and evaluating the input images.

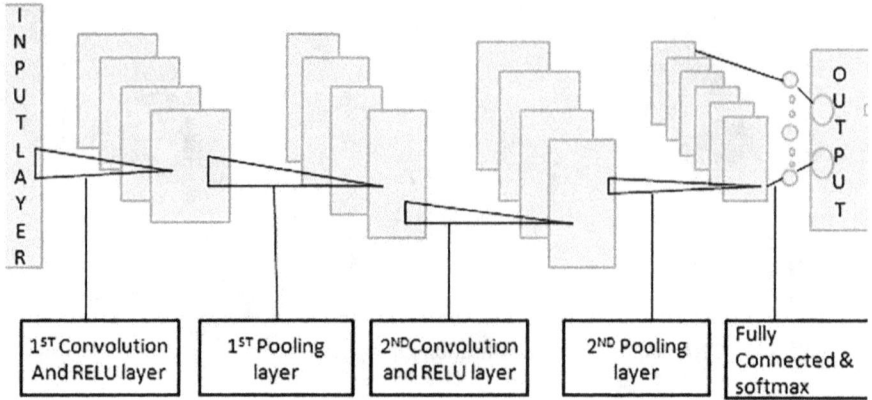

FIGURE 7.4   Layers in CNN.

7.4.3   IMAGE INPUT LAYER

- Input images are of equal dimensions for segregation of normal and ROP-affected images using training and testing process.
- The width and channel dimension of the size is compared with numerical values and verified during the process.
- The channel dimension is 1 for a grayscale picture and the channel dimension is 3 for color pictures.
- Since training the network itself rearranges the information toward the start of training, it can also naturally rearrange the information toward the start of each epoch during training (Figure 7.4).

7.4.4   CONVOLUTION LAYER

- In the convolution layer, the primary argument is filter size, which is the height and width of the channels. The preparation capacity utilized while examining has the dimension 3×3.
- The subsequent argument is the quantity of channels, numFilters, which is the quantity of neurons that interface with a similar area of the information.
- This parameter determines the quantity of highlight maps. The stride and learning rates for this layer can be characterized by utilizing name-esteem pair contentions of convolution2dLayer, along the pictures. In this model, three shows the channel..

7.4.5   ReLU LAYER

- The ReLU is the most common activation function.
- ReLU neglects the negative value and convert it into positive value.

7.4.6 Max Pooling Layer

- Convolution layers(with actuation capacities) are now and then followed by a down-testing activity that decreases the spatial size of the element guide and expels excess spatial data.
- Down-testing makes it possible to expand the quantity of channels in more profound convolution layers without expanding the necessary measure of calculation per layer.
- Maximum pooling is one method of down sampling, which make use of maxPooling2dLayer.
- The maximum pooling layer restores the most extreme estimations of rectangular districts of data sources, indicated by the principal contention, pool size.

7.4.7 Fully Connected Layer

- The convolution and down-inspecting layers are followed by followed by at least one fully connected layer.
- The fully connected layer is where the neurons interface with all of the neurons in the previous layer.
- This layer joins all of the features learned by the past layers over the image to recognize greater models.
- The last fully connected layer combines the features to describe the images. Thus, the output size parameter in the last fully connected layer is identical to the number of classes in the target data.

7.4.8 Softmax Layer

- The output of the fully connected layer is used to standardize the softmax activation function.
- The summed output of the softmax layer can be used to classify the probabilities obtained by the classification layer.
- After the last fully connected layer make a softmax layer to utilize the softmax function.

7.4.9 Classification Layer

- The classification layer is the last layer in the neural network.
- This layer uses the probabilities obtained by the softmax activation function to assign the input to one of the mutually unrelated classes.

7.4.10 Specify Training Options

- The network characterized by the layers is trained using the prepared options and data.
- Table 7.1 demonstrates the accuracy, minibatch loss, and validation loss.

TABLE 7.1

Simulation Results for the Proposed ROP Detection Using CNN

Convolution Layer	Epoch Iteration	Time Elapsed	Mini Batch Loss	Mini Batch Accuracy (%)	Learning Rate
	1–5	66.47–245.67	0.8289–0.1661	61.542–100.00	0.0001
	1–10	31.72–326.17	0.8474–0.0227	53.85–100.00	0.0001
	1–15	33.89–494.55	0.9664–0.0037	61.54–100.00	0.0001
	1–20	43.14–715.00	0.8289–0.0036	61.54–100.00	0.0001
5×5					
	1–5	32.52–164.69	0.8365–0.3571	38.46–76.92	0.0004
	1–10	36.67–360.71	0.5999–0.0152	84.62–100.00	0.0004
	1–15	36.39–496.82	1.5596–0.4981	46.15–92.31	0.0004
	1–20	31.95–617.92	0.61419–0.0001	76.92–100.00	0.0004
	1–5	51.15–218.28	2.3335–0.1172	46.15–100.00	0.0001
	1–10	40.52–471.59	1.5820–0.0263	46.15–100.00	0.0001
	1–15	46.14–699.42	1.6154–0.0010	46.15–100.00	0.0001
	1–20	38.55–783.85	1.1147–0.0000	46.15–100.00	0.0001
6×6					
	1–5	39.02–192.31	0.9035–0.3521	69.23–92.31	0.0004
	1–10	38.88–381.69	0.9466–0.0007	76.92–100.00	0.0004
	1–15	32.07–455.57	1.7594–(–0.0000)	46.15–100.00	0.0004
	1–20	38.22–698.64	3.2383–0.4366	53.85–76.92	0.0004

- Central processing unit (CPU) cannot be used to train a large dataset.
- The cross-entropy loss is a type of loss. The precision level of pictures characterizes the system effectively.

7.4.11 CLASSIFY EVALUATION IMAGES AND CALCULATE ACCURACY

- The labels in the evaluated data is predicted and the final evaluation of the trained network is calculated. Precision is the portion of marks that the system predicts effectively.
- The original data in the evaluation set is matched with the predicted output.

7.5 RESULTS AND DISCUSSION

Figure 7.5 shows the number of iterations versus the time taken for simulation for 5×5 convolutional layer with the learning rates 0.0001 and 0.0004. The results show that learning rate 0.0001 takes more time compared to 0.0004.

Figure 7.6 shows the number of iterations versus time taken for simulation for 6×6 convolutional layer with the learning rates 0.0001 and 0.0004. The results show that learning rate 0.0001 takes more time compared to 0.0004.

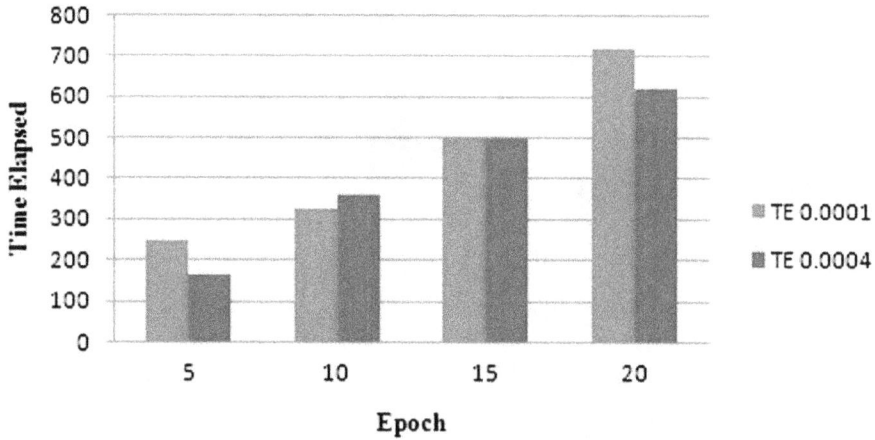

FIGURE 7.5 Epoch versus time elapsed for 5 × 5 convolution layer.

Figures 7.7 and 7.8 show the accuracy of the proposed method for various numbers of iterations with 5 × 5 and 6 × 6 convolutional layer respectively, with the learning rates 0.0001 and 0.0004. The results show that for learning rate 0.0001 accuracy is 100% for any number of epochs and it is varied when the learning rate is 0.0004. From the results it is concluded that the learning rate 0.0001 performs well with greater accuracy than 0.0004.

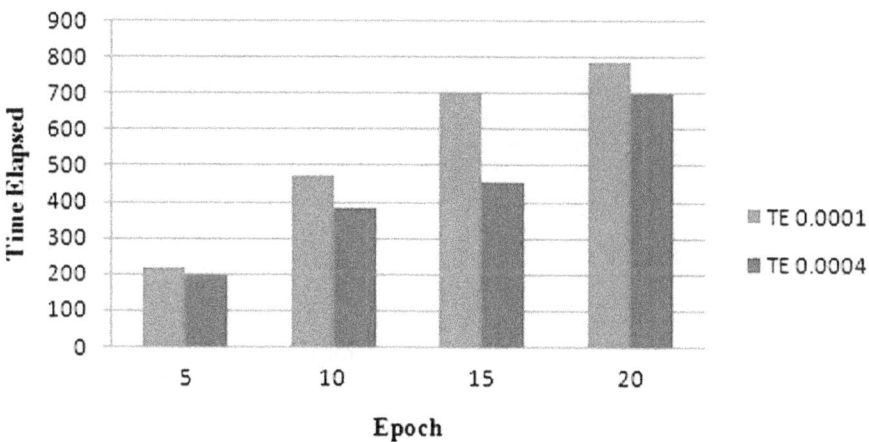

FIGURE 7.6 Epoch vs time elapsed for 6 × 6 convolution layer.

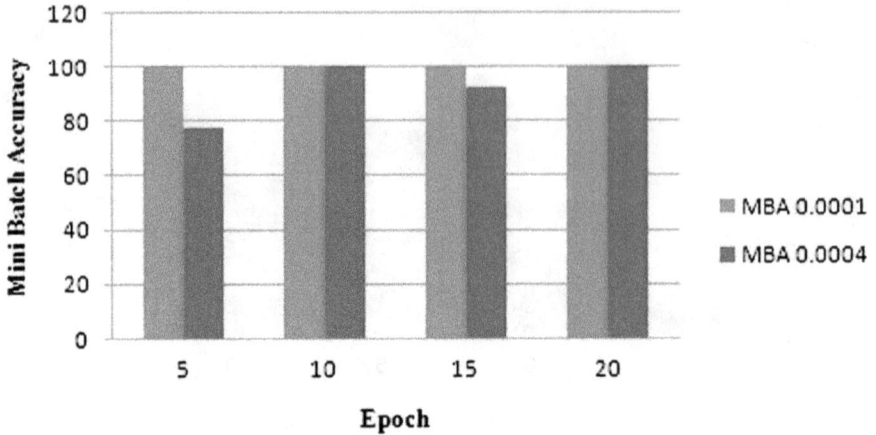

FIGURE 7.7   Epoch vs mini batch accuracy for 5× convolution layer.

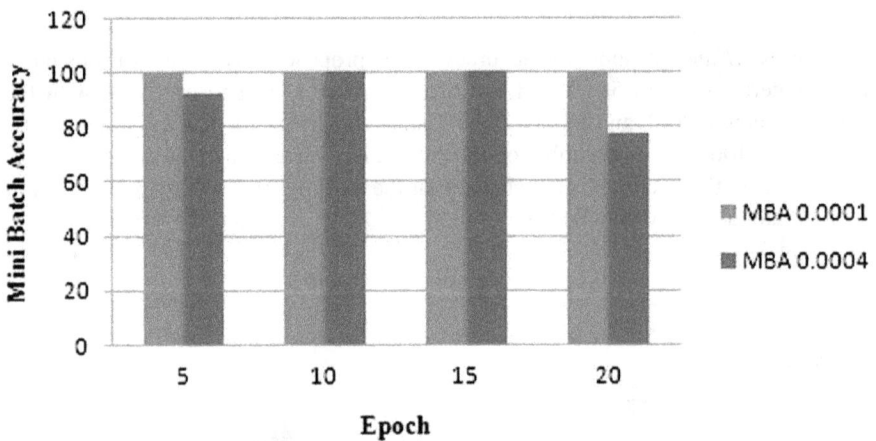

FIGURE 7.8   Epoch vs mini batch accuracy for 6 × 6 convolution layer.

7.6   CONCLUSION

ROP is an optic vessel irregularity that may causes loss of sight if it is not treated properly at the appropriate stage. Detection and diagnosis of ROP can be done in many ways but the proposed CNN method performs well compared to the conventional ANN method. The results obtained from the proposed method show that the detection accuracy will be more when the number of iterations is normalized with a different convolutional layer size. In the proposed method the time taken for detection, batch loss during the process, and accuracy were identified and, based on the results, CNN performs well.

REFERENCES

1. Saunders, R.A., Bluestein, E.C., Sinatra, R.B., Wilson, M.E., Rust, P.F.: The Predictive value of posterior pole vessels in retinopathy of prematurity. J. Pediatr. Ophthalmol. Strabismus32(2), 82–85 (1995)
2. Gole, G.A., Ells, A.L., Katz, X., Holmstrom, G., Fielder, A.R., Capone, Jr., A., Flynn, J.T., Good, W.G., Holmes, J.M., McNamara, J., et al.: The international classification of retinopathy of prematurity revisited. JAMA Ophthalmol. 123(7), 991–999 (2005).
3. Fleck, B.W., Stenson, B.J.: Retinopathy of prematurity and the oxygen conundrum: lessons learned from recent randomized trials. Clin. Perinatol. 40(2), 229–240 (2013).
4. Wilkinson, A., Haines, L., Head, K., Fielder, A., et al.: UK retinopathy of prematurity guideline. Eye (London, England) 23(11), 2137 (2009)
5. Gole, G.A., Ells, A.L., Katz, X., Holmstrom, G., Fielder, A.R., Capone Jr., A., Flynn, J.T., Good, W.G., Holmes, J.M., McNamara, J., et al.: The international classification of retinopathy of prematurity revisited. JAMA Ophthalmol. 123(7), 991–999 (2005)
6. Wilkinson, A., Haines, L., Head, K., Fielder, A., et al.: UK retinopathy of prematurity guideline. Eye (London, England). 23(11), 2137(2009).
7. Fleck, B.W., Stenson, B.J.: Retinopathy of prematurity and the oxygen conundrum: lessons learned from recent randomized trials. Clin. Perinatol. 40(2), 229–240(2013).
8. https://www.mathworks.com/help/deeplearning/examples/create-simple-deep-learning-network-for-classification.html
9. Saha, S.: A comprehensive guide to convolutional neural networks—the ELI5 way. Sumit Saha, December 15 (2018). https://towardsdatascience.com/

8 Impact of Technology on Human Resource Information System and Achieving Business Intelligence in Organizations

Sharanika Dhal, Manas Kumar Pal,
Archana Choudhary, and Mamata Rath
Birla Global University

CONTENTS

8.1 INTRODUCTION

Effective human resource management (HRM) requires all types of advanced technology of the present era. A human resource information system (HRIS) is not new, but it is planned, implemented, and evaluated in the field of human resources (HR) to cope with the changing work culture and technological environment of an organization. With a HRIS it is easy to plan, collect, analyze, and retrieve data as needed by the human resource department. It helps in all types of human resource processes such as human resource planning, recruitment and selection, and training and development. It can be used for performance management and compensation management for recordkeeping of all the facts and figures, i.e., leaves, accidents, superannuation, various employee benefits, etc. The technological environment of the business organization makes it more advanced if it will copes with the external environment. Many researchers describe that integration of the human resource processes with technology is becoming a competitive advantage of the respective industry or organization as compared to others. Having a HRIS is an administrative and strategic advantage to the respective organizations. This chapter is a modest attempt to give a comprehensive analysis to understand the advantages of HRIS in the organization, which may be strategic, competitive, or administrative.

8.2 OBJECTIVE OF THE STUDY

1. To study the concept of the HRIS with its historical evolution.
2. To study the advantages of the HRIS.
3. To study human resource information technology as a competitive strategy and administrative advantage.
4. To study major issues and challenges in the implementation of HRIS.

8.3 RESEARCH METHODOLOGY

An extensive literature survey is done by collecting data from various journals, published articles, online articles, related books, and company websites.

8.4 LITERATURE REVIEW

Firms invest to get leverage or competitive advantages in the form of operational excellence, new product development, customer satisfaction, and improvement in decision-making processes. To achieve this, all management levels concentrate on proper data collection and analysis by using technology in each step of the business process. HRIS is an important part of the whole business process as it helps in each level (Nath, 2015). Even if previously adoption and acceptance of a HRIS was challenging earlier, now it is implemented in most successful industries (Figure 8.1).

HRM is being redefined for better operational performance as it emerges into a new area using new workforce analytics professionals, robot trainers, virtual culture architects, AI integrators, and cyber ecosystem designers. Organizations are using data analytics in HR for prediction.

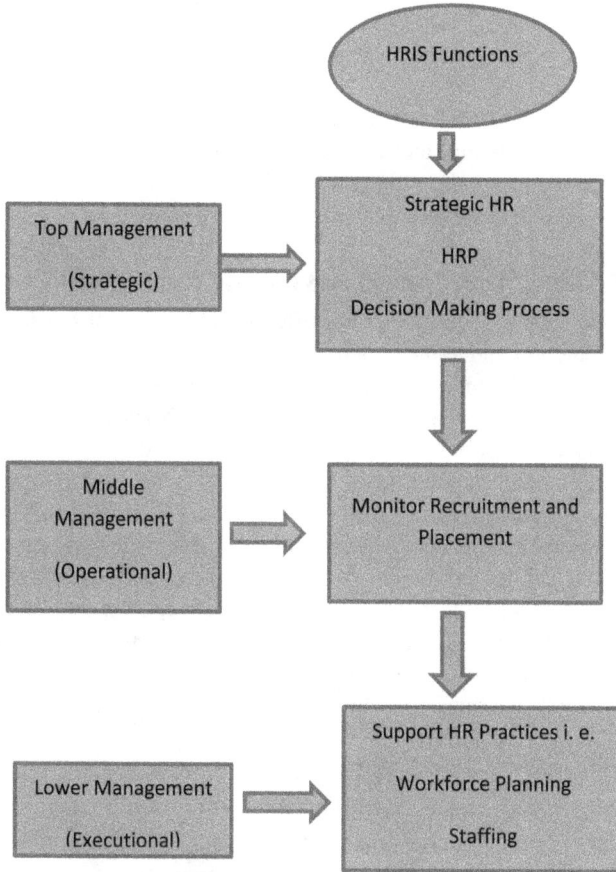

FIGURE 8.1   HRIS functions associated with different levels of HR. (Source: Nath and Naidu, 2015.)

8.4.1   EVOLUTION OF HRIS AND THE CHANGING CONCEPT OF HR INTEGRATION WITH IT

8.4.1.1   Before 1945 (Pre World War II)

HRM was earlier known as personnel management. It had some isolated functions like record keeping, employee relations practices, and child labor issues, which were mainly dealt with by the government. Employee names and addresses were scribbled on 3×5 note cards (Bhuyan, 2014).

8.4.1.2   1945–1960 (Post World War II)

Certain concepts like employee morale, labor utilization and mobilization, job descriptions, extensive reporting to government agencies, and motivation as a social and psychological factor were introduced in the area of HR. Computer technology was used as a facilitator in the HR department (Bhuyan, 2014).

8.4.1.3 1960–1980 (Legislative Era and Emerging HRM)

The personnel department is known as the HR department and computer technology is widely used in various processes. Government and regulatory agencies increase reporting requirements in different areas like occupational health and safety, retirement benefits, tax regulation, and legislative compliance, which requires data collection, analysis, reporting, etc. (Bhuyan, 2014). Enterprise resource planning (ERP) was introduced, which associated with all types of HR functions.

8.4.1.4 1980–1990 (Low-Cost Era and Integration of HR with IT)

The HR department used technology for effective and efficient in-service delivery, through cost reduction and value-added services. HRIS evolves from simple record keeping to different HR functions like recruitment and selection, benefits to management, time management, compensation management, expense reporting, and reimbursement. Some of the definitions of HRIS are as follows:

HRIS is a systematic procedure for collecting, storing, maintaining, retrieving and validating data needed by an organization about its HR, personnel activities and organization unit characteristics (DeSanctis, 1986).

HRIS is a technology-based system used to acquire, store, manipulate, analyze, retrieve and distribute pertinent information regarding an organization's HR consistently.

The human resource department used technology for effectiveness and efficiency in-service delivery, through cost reduction and value-added services. Traditionally it was used only for payroll, and moved from mainframe systems to client-server technology.

8.4.1.5 1990–2000 (Technology Era and the Emergence of Strategic Human Resource Management)

The addition of technology to the organization's product/services and human capital is considered as a competitive advantage and an important part of strategic HRM (Figure 8.2).

8.4.1.6 2000–2010 (Emergence of High Technology and Introduction of Diversified Technological Tools)

According to Scott and Snell (2002), HRM can meet the challenges. Several applications of the HRM system are done. HRMS integrates with human resource planning, preventive and strategic planning, financial planning and risk management, training management and experiences, recruitment and selection process, turnover analysis, attendance reporting and different clerical applications, career and succession planning, compliance with government regulations and analysis, etc. (Hani et al., 2004)

HRIS adoption helps to achieve the following objectives (Ruel et al., 2004):

1. It helps in cost reduction and enhances the innovative characteristics of the employees.
2. It improves the strategic orientation of HRM and facilitates the relationship between management and employees.
3. It helps in client-server improvement and smoothing the functions of HRM.
4. It helps in better compensation management.

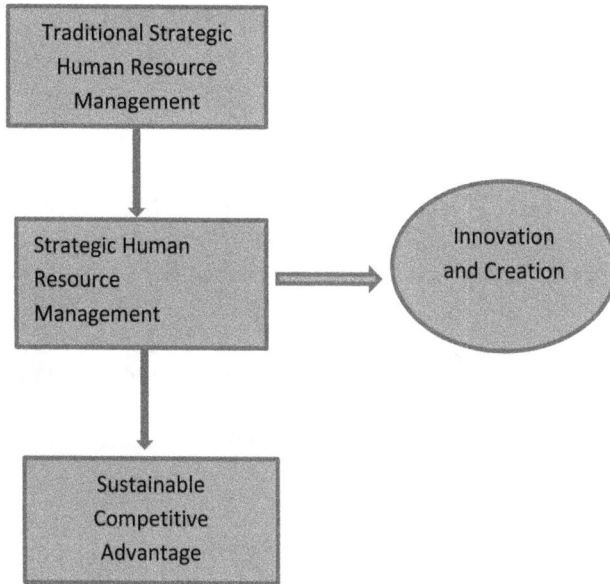

FIGURE 8.2 Traditional strategic human resource management is converted to sustainable competitive advantage. (Source: Bhuyan, 2014.)

In 2007, the term e-HRM was used to designate the action of designing, adopting, and implementing data technology.

8.4.1.7 2010 Onwards (Recent Trends of HRIS)

HRIS facilitates human resource forecasting and planning, provides easy, secure access to data, and helps in effective HR decision-making (Khera, 2012). HRIS is implemented in several business organizations. Currently, enhanced technology is considered a sustainable competitive advantage as a socio-technical system whose purpose is to gather, store, and analyze information regarding an organization's HR department. HRIS comprises computer hardware and applications as well as the people, policies, procedures, and data required to manage the HR function. HRIS integrates with all the processes of HR and leads to error-free results. For example, recruitment is driven by social media and cognitive assessment. Applicant screening systems that use artificial intelligence reduces the man-hours required for recruitment (). Future leaders will require agile thinking, digital skills, global operating skills, interpersonal and proper communication skills, and it will be best if he or she is technologically oriented.

Many noteworthy technologies evolved in this era, such as:

1. Robotics and nanotechnology
2. Network communication technology (broadband and wireless)
3. Convergence technologies(cell phones and PDAs, Internet TVs)
4. Collaborative tools (Web2.0 and portals)

5. Service-oriented architecture (SOA)(e.g., Oracle Fusion and SAP Net Weavers)
6. Business intelligent system
7. Software as a service (SAAS)
8. Infrastructure as a service (IAAS)

There are certain modules that can be included in the HRM system like employee information, recruitment, training, leave, performance management, employee survey, payroll management and attendance, and whether they are implemented depends upon: (Jahan, 2014)

- Effective organizational need analysis and inclusion of proper professionals
- Top management consultation and commitment
- A proper management team
- Effective communication and training

8.4.2 TECHNOLOGY AS A COMPETITIVE STRATEGY

HR can meet the issues and challenges and become more strategic, flexible, cost-efficient, and customer-oriented by leveraging information technology (Scott and Snell, 2002). HRIS facilitates talent management, business process transformation, social networking, time management, HR metrics, workforce planning, ethics, sustainability, safety issues, strategic management, performance management, skill inventory management, legal rules and regulations, and leads to cost awareness, globalization of the workforce, stress management, employee safety, management of workforce diversity, and employee engagement. Firms get competitive advantages in two ways: by accessing special resources or by using common resources efficiently and effectively. Modeling and decision-making depends upon the data collection, transformation, and proper dissemination which will lead the firm to a good strategic position and bring profit (Laudon and Laudon, 2010). There is a great deal of research done on managerial perception toward the effectiveness of human resource information technology on performing administrative and strategic functions of human resource which will work as a competitive advantage for the company. One survey took the feedback of 31 HR managers from private organizations in Bangladesh and suggested that it can be a great competitive advantage to adopt technology in HR department by lowering the cost involved in recruitment and selection, training, and development. By using the HRIS, there is a maximum level of coordination among the various departments and improvement in communication (Islam and Shuvro, 2014). HRIS is a competitive advantage to the organization in public sector organizations as related to the education department (Srivastava, 2018) (Figure 8.3 and Table 8.1).

Many researchers state that the inclusion of information technology or any type of new technique in the product or services will lead to great profitability and create a strategic position in the current market. There are four steps described by Jesuthasen, MD of Wilis Tower Watson, to be followed to implement technology or to do work effectively with an optimal human-machine combination. They are deconstruct, optimize, automate, and reconfigure. Due to the integration of technology with

FIGURE 8.3 Analysis of Michael Porter's competitive forces model. (Source: Laudon and Laudon, 2010.)

business processes like HR, marketing, or operations, globalization of work is possible in this dynamic era.

There are complementary social, managerial, and organizational assets required to optimize the return from information investments described as follows (Figure 8.4):

When calculating investment in HR portals it is seen that the HRIS can be a competitive advantage to the organization (. Table 8.2 shows examples of the administrative advantages of technology.

TABLE 8.1

Five Types of Competitive Forces Are Seen in the Market

FORCES	Traditional Competitors	New Market Entrants	Substitute Products	Customers	Suppliers
Criteria	All firms share market space with others as competitors. Organizations continuously using new technologies to produce new products or to provide services.	New market entrants act as competitors. They hire young workers and implement new technologies in products and services	They lessen the pricing and profit margins. Concentrate on recent trends of technology.	The profitable organization depends to the large measure on its ability to attract and retain customers. There may be little product differentiation.	Suppliers may concentrate on the firm's profits, price, quality of material, delivery schedules, etc.

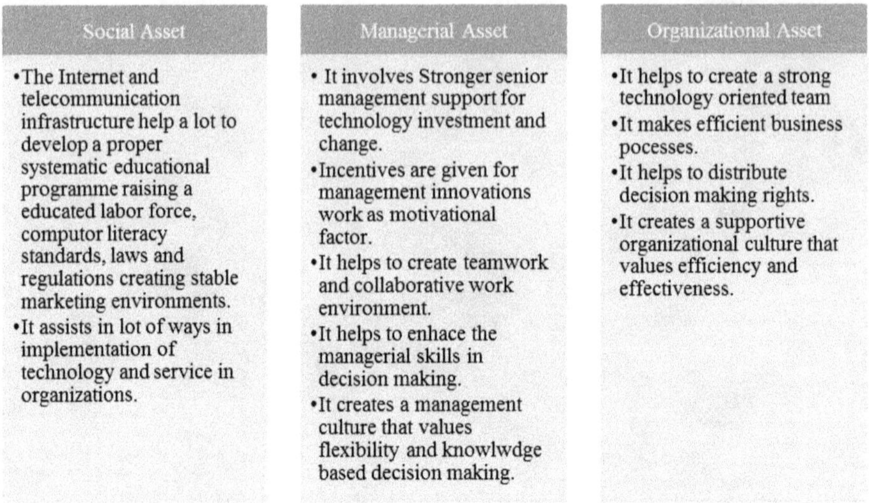

Social Asset	Managerial Asset	Organizational Asset
•The Internet and telecommunication infrastructure help a lot to develop a proper systematic educational programme raising a educated labor force, computor literacy standards, laws and regulations creating stable marketing environments. •It assists in lot of ways in implementation of technology and service in organizations.	• It involves Stronger senior management support for technology investment and change. •Incentives are given for management innovations work as motivational factor. •It helps to create teamwork and collaborative work environment. •It helps to enhace the managerial skills in decision making. •It creates a management culture that values flexibility and knowlwdge based decision making.	•It helps to create a strong technology oriented team •It makes efficient business processes. •It helps to distribute decision making rights. •It creates a supportive organizational culture that values efficiency and effectiveness.

FIGURE 8.4 Complementary social, managerial, and organizational assets. (Source: Laudon and Laudon, 2010.)

TABLE 8.2
Technology as an Administrative Advantage

Serial. No.	Administrative Advantage	Functions
1	Employee Service (ESS)	1. HR data collection 2. Record keeping and access 3. Access by converting client-based service architecture to web-based environment
2	Interactive voice response (IVR)	1. Updating address information 2. Planning for retirement 3. Updating health plan information Reporting on life events like birth of child or life insurance
3	Use of HR portals	1. Different clerical transactions 2. Recruitment and selection
4	Communication	1. Email 2. Read company news and policies 3. Link to job-related information 4. Link of HR data to others
5	Cost reduction in HR functions by calculating different cost)	1. Healthcare costs per employee 2. Pay benefits as a percentage of operating expense 3. Cost per hire 4. Cost for training and development 5. Turnover rates and costs 6. Time required to fill certain jobs 7. Return on human capital investment

When implementing technology in product development or in-service delivery, certain actions are done step by step. For example, there is a feasibility study done which includes all aspects of the current software like the scope of the software, future benefits of the software after implementation, the potential of the software, cost estimation, return on investment calculation, etc. (Jahan, 2014). Some issues and challenges during implementation of HRIS discovered in the literature are shown in Table 8.3.

TABLE 8.3
Issues and Challenges during Implementation of HRIS

Author(s)	Issues and Challenges
Kovach et al. (2002)	1. Technology adaptation
	2. Operation and maintenance issues
	3. Lack of training, following proper process, and end-user involvement
Kumar and Parumasur (2013)	1. Increase in competitiveness in collecting data
	2. Reengineering of business processes
	3. Difficulties in adapting
	4. Long implementation period
	5. Acceptance of new or upgraded HRIS
	6. Management of change processes
	7. Lack of understanding
	8. Insufficient management commitment
	9. Fear of change
	10. Result in jobless/altered leave entitlement/shift arrangement
	11. Up-to-date information
Sadiq et al. (2012)	1. Lack of cooperation
Weeks (2013)	1. Lack of strategic/operational efficiency, user-friendly interface, and commitment
	2. Insufficiency in integration with other systems within the organization
	3. Complicated system
	4. Inflexibility
	5. Insufficient fund allocation
Jahan (2014)	1. Expensive
	2. Lack of end-user participation
	3. Threatening and inconvenient
	4. Lack of change management
Nath (2015)	1. Process improvement
	2. Employee acceptance
	3. Security issues
	4. Cost of operation
	5. Adaptation challenges
	6. Technical limitation
	7. Responsibility of HR department
	8. Resistance to change
	9. Software errors and bugs
	10. Change in workflow

(Continued)

TABLE 8.3 (*Continued*)
Issues and Challenges during Implementation of HRIS

Author(s)	Issues and Challenges
Ali (2017)	1. Lack of funds, expertise, cooperation, and professional staff
	2. Inadequate knowledge
	3. Network problems
	4. Technical problems
	5. Time consumption

All the issues and challenges can be overcome by proper training and development, effective training need analysis, top management commitment, effective communication, the inclusion of key people, a dedicated team or professional team (Jahan, 2014), and effective and accurate decision-making (Islam and Shuvro, 2014). Implementation of HRIS helps in cost reduction, competency management, knowledge management, the motivation of employees, effective communication, proper monitoring and feedback, and adaptability and flexibility (Nath, 2015). It also has other benefits like reducing chapter work, data improvement, quick response, easy access to information, service improvement, reducing errors, enhancing competitiveness, time-saving, etc. (Ali, 2017).

8.5 IMPLICATIONS

This chapter highlights the strategic importance of HRIS implementation, issues, and challenges faced by organization when implementing HRIS in the organizational process. It also gives insight to many of the researchers, academicians, IT professionals, organizations, and others who are associated with the IT sectors as it includes some of the related facts regarding the HRIS.

8.6 SCOPE OF THE STUDY

Future studies can be done on the impact of HRIS on organizations or HR practices at all levels of employees or in all the sectors by investigating the public and private sectors differently. Alternatively, the study can be done by taking different areas like IT and non-IT sectors. Issues related to the implementation of HRIS or longitudinal research can be done to study the effectiveness of HRIS in a particular situation. There can be studies on the adaptation of technology in HR practices in a diversified workforce with consideration of internal and external factors.

8.7 CONCLUSION

Change management and an increase in employee morale may be the key factors to encourage organizations to adopt the technology with the organizational functions. Proper training will be helpful for the successful implementation of the HRIS.

The future workplace will need a huge usage of technology and analytical skills. That's why organizations need proper analysis of training needs and feedback about technological adaptation. There is no doubt that if there will be a collective effort between the organization and employees to run the technology smoothly for all the processes, then all the apprehensions relating to the technology will decrease and a day will come when man and machine will complement each other and give efficient and effective results.

REFERENCES

Ali, H. A. (2017). Human resource information systems. *International Journal of Management and Applied Science (IJMAS)*, *3*(12), 43–45.

Bhuyan, F. M. (2014). Historical evolution of human resource information system (HRIS): An interface between HR and computer. *Human Resource Management Research*, *4*(4), 75–80.

DeSanctis, G. (1986). Human resource information systems: A current assessment. MIS Quarterly, *10*, 15–27. http://dx.doi.org/10.2307/248875

Islam, M. F., and Shuvro, R. A. (2014, December). Administrative and strategic advantages of HRIS: An exploratory study in the private sector of Bangladesh. *Stanford Journal of Business Studies*, *6*(1), 65–73.

Jahan, S. (2014, May). Human resources information system: A theoretical perspective. *Journal of HR and Sustainability Studies*, *2*, 33–39.

Khera, M. K. (2012, Oct). Human resource information system and its impact on human resource planning: A perceptual analysis of information technology companies. *Journal of Business and Management*, *3*(6), 6–13.

Kovach, K. A., Hughes, A. A., Fagan, P., Maggitti, P. G. (2002). Administrative and strategic advantages of HRIS. Wiley Online Library. https:/ /doi.org /10.1002/ert.10039

Kumar, N. A., and Parumasur, S. B. (2013). The impact of HRIS on organizational efficiency: Random or integrated or holistic. *Corporate Ownership & Control*, *11*(1), 567–575. http://dx.doi.org/10.22495/cocv11i1c6art4

Laudon, K. C., and Laudon, J. P. (2010). *Information System in Global Business Today* (11th ed.). New York: Pearson.

Naidu, P. S. (2015). HRIS Efficiency and its impact on Organization. *International research Journal of Management science and Technology (IRJMST)*, *6*(7), 85–98.

Nath, P. S., and Naidu, J. G. (2015). HRIS efficiency and its impact on organization. *International Research Journal of Management Science and Technology (IRJMST)*, *6*(7), ISSN-2250-1959

Papia, S., and Nath, D. G. (2015). HRIS efficiency and its impact on the organization. *International Research Journal of Management Science and Technology*, *6*(7), 85–98.

Ruel, H. J. M., Bondarouk, T., and Looise, J. C. (2004). E-HRM: Innovation or Irrigation. An explorative empirical study in five large companies in web based HRM. *Management Revue*, *15*(3), 364–380.

Sadiq, U. (2012). The impact of the information system on the performance of HRD. *Journal of Business Studies Quaterly*, *3*(4), 77–91.

Scott, A., and Snell, D. S. (2002). Virtual HR Departments: Getting out of the middle. *Working Paper series (1-8), Centre for Advanced Human Resource studies (CAHRS)*.

Srivastava, P. (2018). Comparative analysis of effectiveness between E-HRM and traditional HRM. *International Journal of Advance Research, Ideas and Innovations in Technology*, *4*(6), 1–5.

Weeks, K. O. (2013). An analysis of human resource information systems impact on employees. *Journal of Management Policy and Practice*, *14*(3).

9 Proficient Prediction of Acute Lymphoblastic Leukemia Using Machine Learning Algorithm

M. Sangeetha, K.N. Apinaya Prethi, and S. Nithya
Coimbatore Institute of Technology

CONTENTS

9.1　INTRODUCTION

An artificial neural network is a system that is similar to the human brain in terms of structure and information processing. It is a simple model of the brain that deals with linear and nonlinear relationships between the input and output. A neural network is a collection of neurons, which is a mathematical function that gathers and processes data per the system design. The neural network can be used for any number of tasks, but it became popular for classification because of its precision. For example, face recognition can be used by our mobile devices to securing a phone from intruders. A neural network is trained with a set of images and during validation it finds out how closely a captured image matches stored

135

images (trained data). Back-propagation plays a major role in neural network, by which high accuracy is achieved. In the back-propagation process, feedback is given to a neural network and the achieved output is compared with the output which was meant to be produced and the weighting of the neurons adjusted accordingly. This technique helps a neural network to achieve high accuracy in a shorter period than any other traditional system [1]. This makes neural networks ideal for speech recognition, text classification, language processing and semantic analysis.

Machine learning (ML) [2] provides many algorithms to process, discover, identify, suggest and predict future actions. ML techniques may become coercion when the entire world starts producing big data streams in daily life through social media applications like WhatsApp, Facebook, Instagram, etc. [3]. However, ML algorithms have to overcome several challenges when they works on data which are huge in size [4]. Data storage and manipulation will be major challenges which can be resolved by parallelizing the process [5]. In cloud computing, server may go down due to huge amounts of data being processed at a same time. Data storage and security is also a major issue. This will lead to many problems like higher response time and high bandwidth consumption. To overcome this, data processing is being pulled closer to the user with edge computing. Edges may be a gateway, router or any network device that is located near the user and responds to the user in a short period of time. The response will get stored in the cloud if needed for further prediction processes. Edge computing also has many challenges, such as connectivity, power management and security.

Support vector machine (SVM) is one of the most popular ML algorithms, which can be used for classification and regression. In classification, SVM works faster and results in high exactness when it deals with limited amounts of data. As a result, it creates a hyper plane between two different classes (group of data). SVM has compensations when solving small model, nonlinear and high dimensional classification problems [6].

A clear understanding of the concepts of artificial intelligence, ML and deep learning is very important to have a strong foundation so that ground-breaking ideas can proceed. All three are interrelated and their intention is similar. Artificial intelligence will have a program to learn and respond like a human. Artificial intelligence includes any techniques like ML algorithms and neural networks which either enable the machine to behave like a human or teach the computer to answer smartly. The ML algorithms help the machine to assess the behavior of the user and the input. A machine learns by using a certain amount of data and then being tested. Artificial neural networks developed with three main layers: one input layer, one hidden layer and one yield layer. Each layer has a number of nodes and information will flow to another node.

Clustering is grouping objects which are similar in any one of their characteristics. Clustering is key to research and it is widely used in many fields such as marketing, insurance, medicine, digital image processing, knowledge discovery and data mining. In this project, our input will be images, hence k-means clustering will be an image classification algorithm. Many clustering algorithms are available for image classification like k-means, k-medoids and hierarchical clustering. Clustering can be done in two ways: hard and soft clustering. In hard clustering, data points will be in any one cluster. A data point will not exist in two clusters at the same time. Data points will be the member of the cluster and not a member of other clusters. Soft clustering finds the probability of data points to be members of a particular cluster. Each data points will not exist in separate cluster as like hard clustering.

Currently, different methods are used to form clusters. The types of clustering methods are connectivity clustering, centroid clustering, distribution clustering, density clustering and hierarchical clustering. Data points with less distance between them are formed as clusters in connectivity clustering. As the name suggests, connectivity between data points decides the cluster members. Selecting the required centroid and grouping the nearest data points form the cluster in centroid clustering. Distribution clustering finds the probability that all data in the cluster will be appropriate to the identical distribution. When the data points satisfy the particular condition that defines the cluster center then they join the cluster. DBSCAN is one of the most popular examples for density-based model. Hierarchical clustering develops a tree-shaped cluster.

k-Means clustering is the simplest centroid-based clustering and unsupervised learning algorithm that is widely used. In the k-means algorithm, k indicates the number of centroids and it will be chosen as the first step of the algorithm. Each data point will get added to the cluster that is nearest. This algorithm will work iteratively and create a non-overlapping, unique group, i.e., cluster [7, 8]. k-Means follows the expectation and minimization approach to resolve the problem. The process of assigning data points to a cluster is called the e-step. The process of finding k-centroids is called the m-step.

The steps of the conventional k-means algorithm are as follows:

1. Indicate the total clusters and pick out the k number of points as initial centroids accordingly.
2. Form k clusters by allocating all points to the closest centroid.
3. Recompute the centroid of each cluster.
4. Again assign the data points to the nearest cluster.
5. Repeat steps 3 and 4 until no changes are needed in the centroid.

An image is a collection of a large number of pixels. Objects in the image get classified by their edges and curves For example, an autonomous car is an automatic car that detects the objects before it using image classification.

Image segmentation is the process of separating object from their background and then uncovers the needed targets. Watershed segmentation is based on mathematical morphology and thus it overcomes many shortcomings faced by other image segmentation methods [9]. This method is faster and accurate. A watershed is the region that collects rainfall that drains into a river. This segmentation method decomposes the image pixel either into ridges or valleys. If the surface has high intensity it will be a ridge, whereas the surface with low intensity will form a valley. Each isolated valley will have different colored water. Water from the different isolated valleys will merge when the water level increases. Barriers should be built in the location where valley water gets merge to avoid the merging process. The process of increasing the water level and building the barrier will be done until all the peaks gets submerged in the water completely. Finally, the barrier built in the merging place will give the accurate segmentation result. Oversegmentation can occur when noise and irregularities in the input image gets decomposed into a large number of regions. This problem can be rectified by having a constraint to decompose pixels into region [10, 11].

To proceed we will analyze the above-mentioned ML algorithms with a project of discovering leukemia in blood smear images.

9.2 PROBLEM SOLVING

9.2.1 PROBLEM STATEMENT

Leukemia is a hazarding and fatal disease which affects many people, especially children [12]. This disease does not usually have any symptoms. But detecting the presence of the disease at an early stage will allow for earlier recovery. To find the leukemia, blood sample images can be taken as input. A data mining technique that works efficiently on huge amounts of data may be applied.

9.2.1.1 Solution

Leukemia is detected by using image processing and ML techniques [13]. The input image for this project is a blood smear microscopic image. The input image undergoes image preprocessing and enhanced, which helps in visual interpretation and understanding the image. The image is first de-noised and resized. The image resizing is used to scale the image to a standard size. The noise removal from the image retains the original image. This image is clustered using k-means and segmented using the watershed algorithm. Further preprocessing techniques include grayscale conversion and binarization. The clustering algorithm is used to cluster the leucocytes. Then the leucocytes are extracted separately by watershed segmentation and feature extraction technique. Therefore, the white blood cells (WBCs) are separated from the original blood smear image. The classifier algorithm is used to detect the cell as cancerous or non-cancerous (Figure 9.1).

The input images are of two types: training set and testing set. The training set implies the set of saved images in the system to train the classifier. The test set implies the set of images which is provided by the user for detecting the disease. The input image is subjected to various processing steps. This project get an image as input and preprocesses it followed by clustering, segmentation, feature extraction and classification. The preprocessing steps includes resizing the image to the required dimension. Then, the image is subjected to noise while capturing. Hence, the noise is removed from the image. The image is segmented and hence the overlapping cells are separated. The clustering algorithm

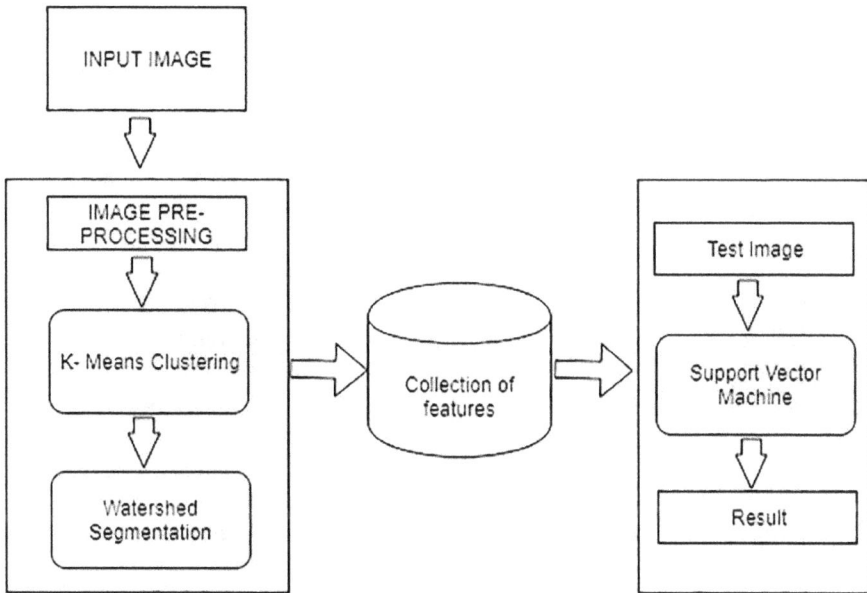

FIGURE 9.1   Steps to be followed to detect leukemia.

used to group leucocytes is the k-means clustering algorithm where the image is partitioned into k clusters. The extracted clustered image is then converted to grayscale image. The clustered image is again segmented using the watershed segmentation algorithm. The Canny edge detection algorithm is used to detect the edges of leucocytes. The features like area, perimeter and compactness are calculated and are stored in .csv file for future analysis. The area of WBCs is given by numbering the count of white pixels in the binary image. The SVM classifier is used to classify the image as cancerous or non-cancerous and thus the result is obtained.

9.2.2   System Design and Implementation

In image processing, the input must be an image. Hence, the input for this project is a blood smear microscopic image. A blood smear is a thin layer of blood smeared on a slide and stained for examination under a microscope. Some other names of blood smear are blood cell morphology, blood film, peripheral blood film and peripheral smear. For verification of a blood smear image, laboratory experts focus on a problem that may not be identified by computer analysis.

This image will undergo various stages such as preprocessing, clustering and classification to find either cancerous image or not.

9.2.2.1   Image Preprocessing

The original image undergoes a sequence of steps in preprocessing to undergo further advanced process such as clustering and classification.

Procedure

Step 1: Read the Input image.
Step 2: Image resizing.
Step 3: Image denoising.

Step 1: Read the Input image

The input image stored in the system can be retrieved by specifying the location where it is saved. The image can in retrieved either through OpenCV package or Matplotlib package. The main difference between OpenCV and Matplotlib is the mode of color projection. OpenCV is in BGR mode whereas Matplotlib is in RGB mode. Hence, color images will not be displayed correctly in Matplotlib if the image is read with OpenCV. Hence, the OpenCV package is used to read and display the image. The image is converted to nd_array format using Numpy. nd_ array is collection of similar elements which can be accessed through a zero based index. Elements of the nd_array will take uniform sizes of memory. Slicing is a technique which can be used to extract the nd_ array objects. The input image retrieved through OpenCV can be utilized for further manipulation (Figure 9.2).

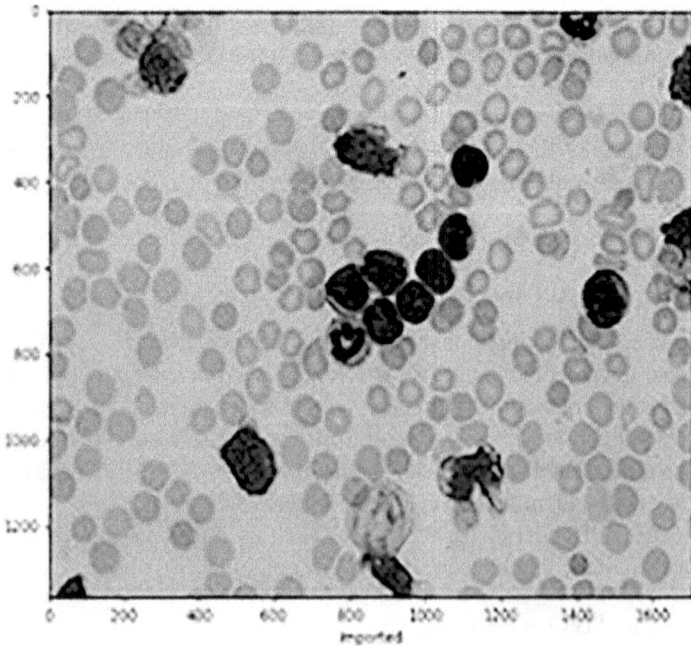

FIGURE 9.2 Input blood smear image.

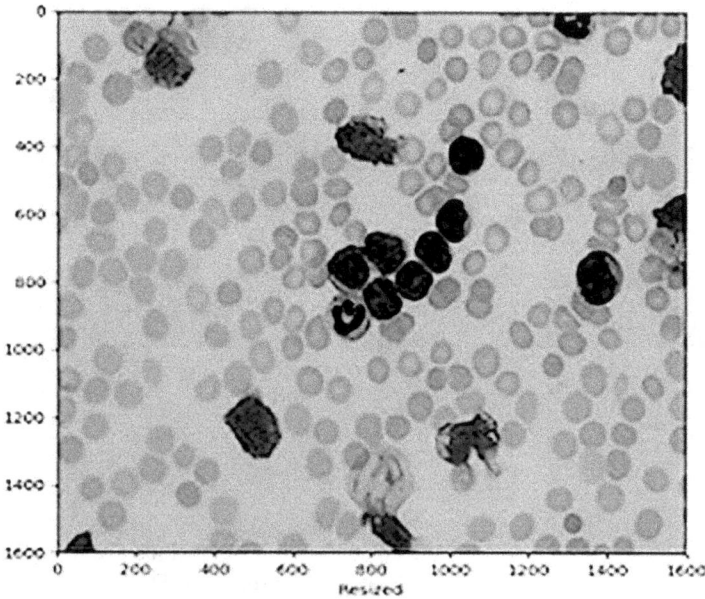

FIGURE 9.3   Resized image.

Step 2: Image Resizing

Image resizing is a simple and import step in preprocessing. The image may be captured through many mediums such as a digital camera, mobile camera, etc.; different images have different dimensions. The main reason for resizing is to establish a base size for all images. The standard size defined is (1600, 1600) where the value represents row and column size of the nd_array. Here, it performs interpolation to upsize or downsize images to acquire the standard resolution (Figure 9.3).

Step 3: Image Denoising

Denoising is the process of removing unwanted noise using smoothing techniques such as Gaussian smoothing. Presence of noise may lead to inaccurate results. In OpenCV, the module cv2.fastNlMeansDenoisingColoured () works with the colored image. The noise is filtered from the input image. The noise reduction is very important since it provides accurate results after analysis of the image. The presence of noise may cause a drastic change in the result from the expected result (Figure 9.4).

9.2.2.2   *k*-Means Clustering

k-Mean is one of the most promising unsupervised clustering algorithms to partition the dataset into *k* clusters, in which each data point belongs to nearest mean. The proposed architecture uses the *k*-means clustering algorithm to cluster WBCs. The best *k*-means clustering will reduce the total intercluster variance. It starts with the promising procedure of classifying an input image to a number of defined *k* clusters.

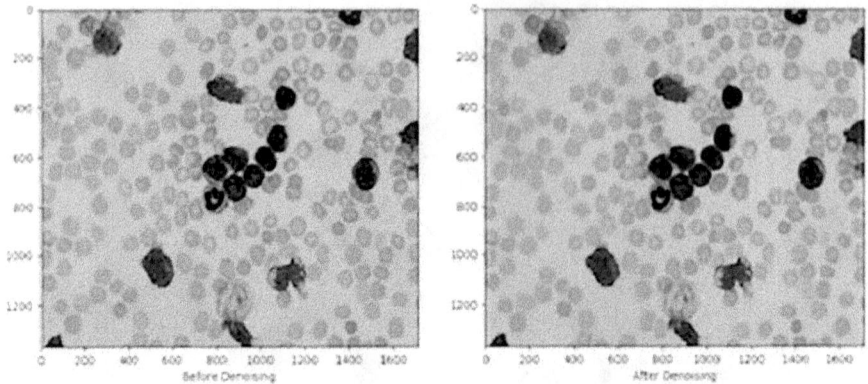

FIGURE 9.4 Denoised image.

Then the process starts over all the clusters and observations are made, the same process will continue until a favorable result is reached. During the observation data points are mapped to any nearby cluster then the dataset is tuned until the final result.

The algorithm

Step 1: Define number of k, which is used to cluster the given data set.

Step 2: Identify k-point as cluster center (k-centroid)

Step 3: Each object or data point is grouped into the nearest k.

Step 4: After proper clustering, the places of the k-centroids are recalculated.

Step 5: Steps 3 and 4 are repeated until the positions of the centroids no longer move.

The module named k-means () from cv2 package is used for image clustering. The various inputs which are provided to the algorithm are as follows:

Inputs

Step 1: Input image data type.

Step 2: Number of clusters – the number of groups required after partitioning.

Step 3: Iteration termination criteria

Step 4: Number of attempts – the number of times algorithm is executed.

Step 5: Initializing centroids

Step 1: Input image data type

The input image must be an nd_array with float 32 data type.

Step 2: Number of Clusters

The number of groups required after partitioning must be specified. The different clusters can be accessed by using label variable.

Step 3: Iteration Termination Criteria

There are three criteria to terminate the iteration: maximum iteration, epsilon and type, which are explained as follows:

9.2.2.2.1 *Maximum iteration*

The maximum number of iterations to terminate the algorithm is specified. Here, the maximum iteration is ten to get an efficient result. When the iteration reaches ten, the algorithm is terminated.

9.2.2.2.2 *Epsilon*

The accuracy required to be reached is specified as epsilon. Here, the specified epsilon value is 1. When the accuracy reaches unity, the algorithm is terminated.

9.2.2.2.3 *Type*

TERM_CRITERIA_EPSILON+MAX_ITER – The iteration terminates if the specified accuracy (epsilon) value or the maximum iteration is reached. Hence, when any one of the above conditions is met the algorithm stops.

Step 4: Number of Attempts

The number of attempts is used to specify the number of times the algorithm is executed using different initial labeling. The algorithm returns the label that yields the desired output.

Step 5: Initializing Centroids

A flag is used to specify how the initial centers are taken. Here, random centers are chosen and the above procedure is followed. Figure 9.5 shows the output.

9.2.2.3   Watershed Segmentation

Any grayscale image can be observed as a topographic surface that contains high intensity (denoting peaks/hills) and low intensity (denoting valleys).First pack every valley (local minima) with different labels (colored water). Based on peaks (gradients) and valleys nearby, the water level will colliding with each other which injures the image. To prevent this, build barricades in the locations where water merges. Repeat the action of filling water and making barricades until all the peaks are under water. The barricades you created give you the segmentation result. Oversegmentation will result due to noise or any other irregularities in the image. The marker-based watershed algorithm was implemented by OpenCV to denote the valley points to be merged and not to be merged.

The algorithm

1. Identify the region to be a foreground or object with one color.
2. Identify the region to be a background or non-object with another color.
3. Finally, label the regions neither foreground nor background with 0.
4. Apply watershed algorithm.
5. The marker will be updated and the boundaries of objects will have a value of –1.

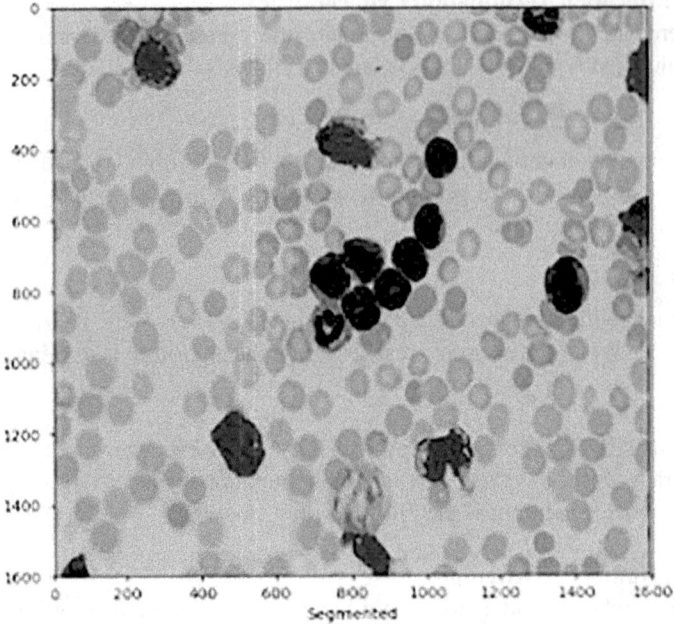

FIGURE 9.5   Segmented image.

Inputs

Step 1: Input image data type.
Step 2: Label foreground image.
Step 3: Label background image.
Step 4: Label unknown region.
Step 5: Apply watershed algorithm.

Step 1: Input image data type
   The colored blood smear image will give as input image. Then it will be converted to a grayscale image and binarization applied to it to get a binary image from the grayscale image.
   Color image → grayscale image → binary image.

Step 2: Label Foreground Region
   Label those region or components of the image which are sure of being foreground. An approximate estimate of the image is found using Otsu's binarization. This is used to automate image thresholding in the sense of separation of foreground image from background image. Morphological closing is used to remove any small holes in the object. Hence the foreground image is obtained (Figure 9.6).

Step 3: Label Background Region
   Dilation adds pixels to the boundary to increase the object boundary to the background. Hence the resultant image is a background, since the boundary region is removed (Figure 9.7).

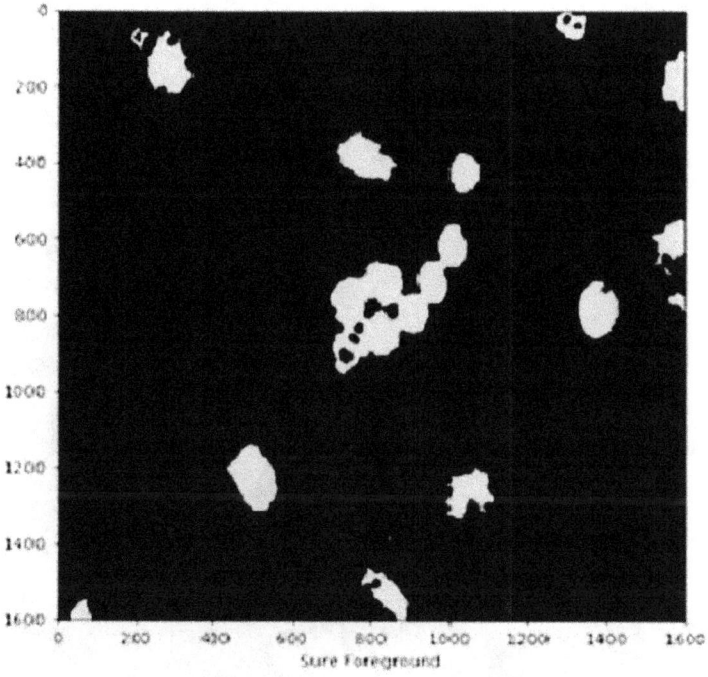

FIGURE 9.6   Foreground image.

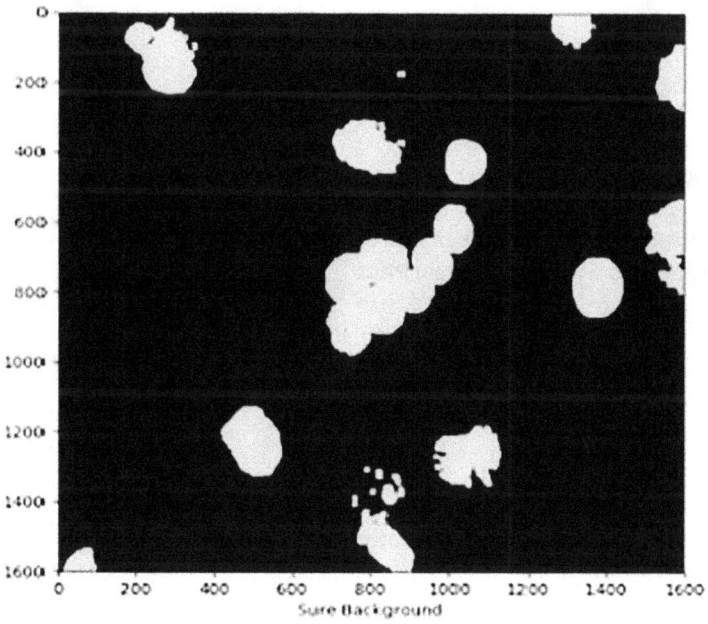

FIGURE 9.7   Background image.

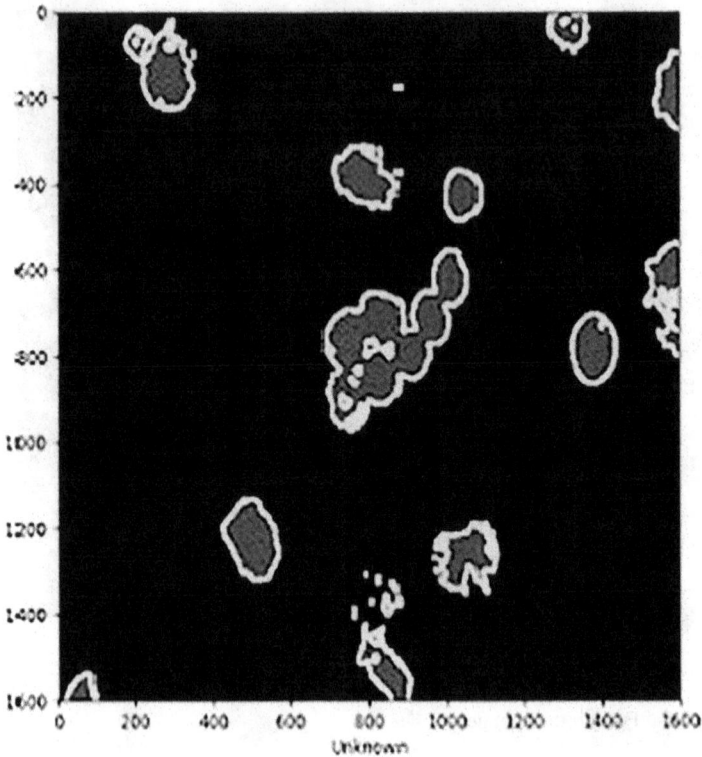

FIGURE 9.8   Unknown region of image.

Step 4: Label Unknown Region

The region that is neither background nor foreground will be declared as the unknown region. can be obtained by subtracting the sure foreground region from the sure background region (Figure 9.8).

Step 5: Apply Watershed Algorithm

Markers are arrays of same size as that of original image, but with int32 data type. Sure foreground and background regions are marked positive. Unknown region is marked with 0. When watershed algorithm is applied, boundary will be obtained with marker value as −1.

9.2.2.4   Feature Extraction

Feature extraction is the method of extracting elements of interest from an image and optimizing them for supplementary processing. In feature extraction the raw data in an image will be converted into more meaningful data on a desired point. The resulting representation can be subsequently used as an input to the classification techniques, which will then label, classify or recognize the semantic contents of the image or its objects.

Procedure

Step 1: Grayscale conversion
Step 2: Binarization of image
Step 3: Edge detection
Step 4: Extracting features

Step 1: Grayscale Conversion (Color image-→ grayscale image)
   The input colored image is represented in a RGB plane. The given image (x,y,3) where 3 represents RGB, is converted to (x,y,1) where 1 represents gray color using np.dot() function. This returns the dot product of (x,y,3) and (x,y,1). To get the grayscale image, the red plane is dot product with the constant 0.299, the green plane is dot product with 0.587 and the blue plane is dot product with 0.114 (Figure 9.9).
Step 2: Binarization of image (grayscale image → binary image)
   The algorithm used for converting a grayscale image (pixel image) to binary image is Otsu thresholding. The steps involved are
   i. Covert into grayscale
   ii. Apply threshold value. It can be either fixed or adoptive (in our case the threshold value is defined as 255)

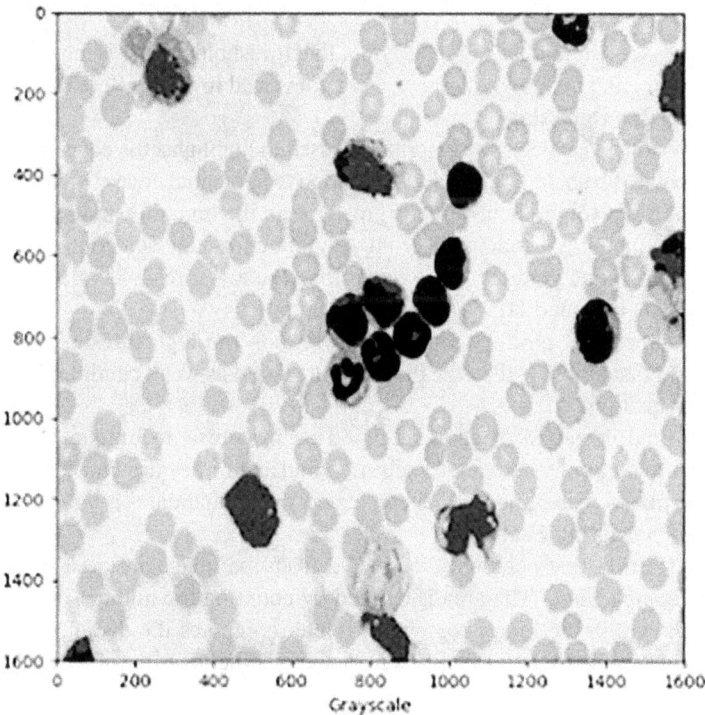

FIGURE 9.9 Grayscale image.

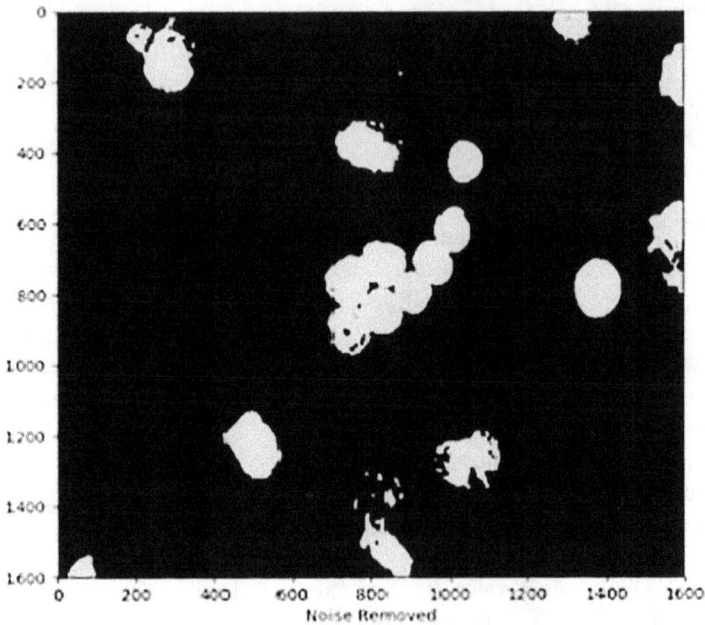

FIGURE 9.10   Binarized image.

The pixel value that is underlying the threshold value is converted to 0 (white) and the overlying values are converted to 1 (black) (Figure 9.10).

Step 3: Edge Detection

The Canny edge detection algorithm is used to highlights the edges of WBCs. A double threshold is applied to determine the potential edges (Figure 9.11).

- If the pixel gradient value is greater than the high threshold value, then the pixel is considered a strong candidate for edge detection.
- If the pixel gradient value is less than the low threshold value, then the pixel is turned off.
- If the pixel gradient value is in between the high threshold and low threshold value, then the pixel is considered as weak candidate for edge detection.

The final detection of edges is based on hysteresis technique. When the weak and strong candidate pixels are connected, they are considered to be edge pixels. The remaining unconnected weak candidates are turned off.

Step 4: Extracting Features

The features that need to be extracted from the image are area, perimeter and compactness. The area is defined by counting the number of non-zero pixels within the image region. Perimeter computes the distance between consecutive boundary pixels. Compactness is the measure of a nucleus.

$$\text{Compactness} = \frac{\text{Perimeter}^2}{\text{Area}}$$

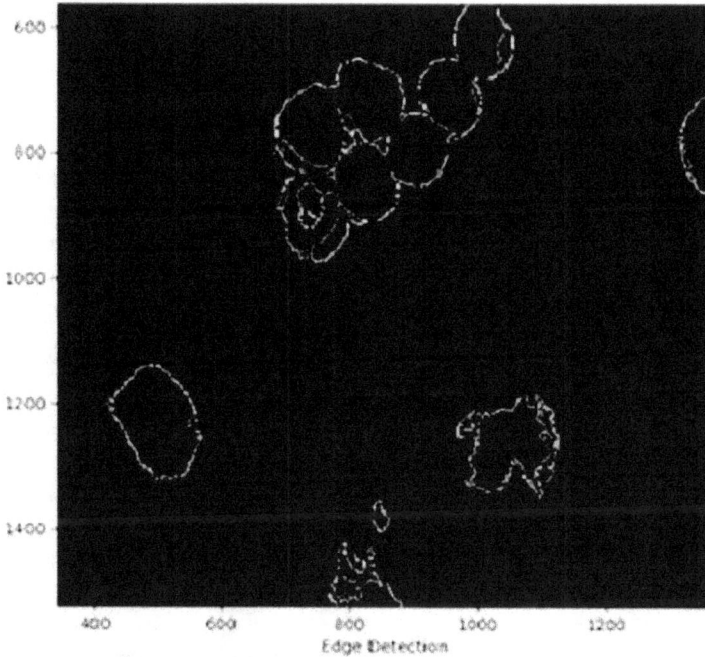

FIGURE 9.11 Edge detection.

9.2.2.5 SVM Classifier

SVM is a well-organized technique that supervises the entire process based on which classification need to be done. In this ML technique, supervision is conceived by the training set. The proposed system will learn from the training set and act accordingly. In ML, two validating strategies are used for classifying training and test data: hold-out and cross-validation. In hold-out, the input dataset will get classified as training and testing data whereas in cross-validation it will get classified into k groups. One group is used as a training set and the remaining group is used as a testing set. Each group has to be used as test data; only then is the process complete. In our proposal we have used the hold-out strategy such that input data is classified as either the training or testing set. Providing that the group of training examples each belong to one of two classes, the SVM model classifies the training example by plotting a hyper plane. The function of a kernel is to get the input and convert it into an appropriate form like a graph with a hyperplane. This kernel function may be linear, nonlinear, RBF or Gaussian. We could find many methods of finding a hyperplane that may classify the data. A wise way to choose a hyperplane is the one that represents the maximum separation between two classes. The most common kernel function is RBF because of its low response time.

Input Parameters

Step 1: Input: Hold-out strategy
Step 2: Score function

Step 3: Loss function
Step 4: Weight matrix

Step 1: Input: Hold-out strategy

The input training test contains data which is given as input to the proposed system and from which the proposed system will learn how to react to the input data. The input testing data is given to the proposed system after the proposed system gets trained from the training set. Mostly of the dataset used as training set (80%) and 20% of the dataset is used as a testing set. The training dataset can be chosen from the input dataset either randomly or the first 80% data.

Step 2: Score function

We need to convert data into class labels for further processing. This conversion is done by the score function which applies score function f (the score function) on the input and yields the predicted class labels as output.

Step 3: Loss function

A loss function computes whether the predicted class labels approve with our ground-truth labels. Lowering the loss and increasing the accuracy have to be attained at least on the training data. If this is also accomplished in the test data then the accuracy level of the system will be high.

Step 4: Weight matrix

The output of our score function and loss function will determine the values of the weight matrix, which is denoted as W. The values in the matrix will be the parameters of the classifier and it will fine-tuning with the score and loss function's value in order to maximize the classification accuracy.

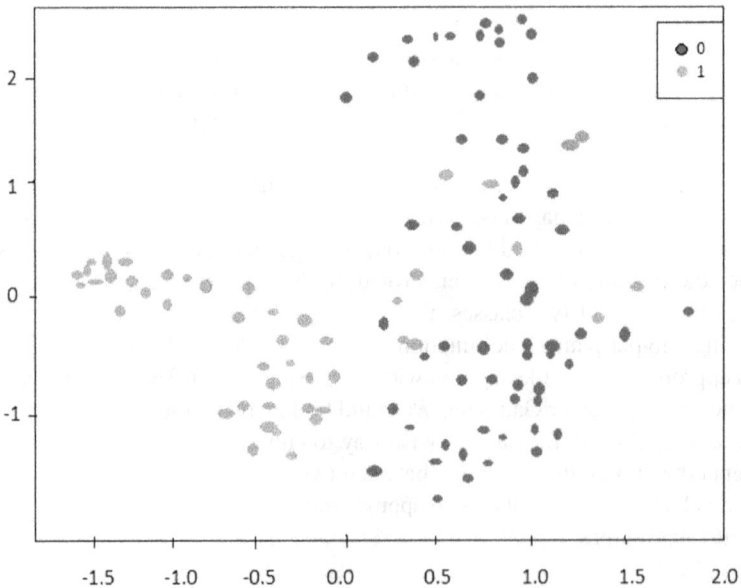

FIGURE 9.12   Plotting features.

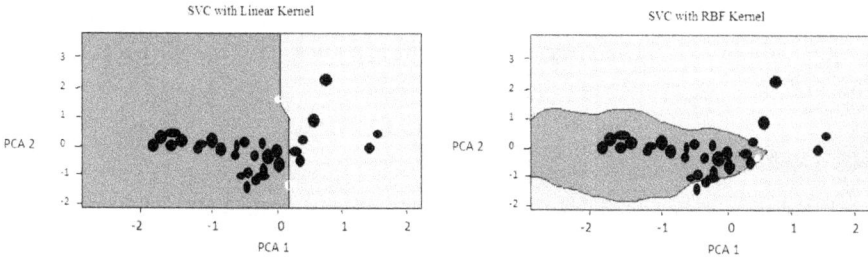

FIGURE 9.13   PCA graph for linear and RBF kernel.

9.2.2.6   PCA Graph for Linear and RBF Kernel

Principal component analysis (PCA) is an important dimension-lessening technique which reduces a large set to a small one but still holds the information of the large set(Figure 9.12).

SVM deals with linear and RBF kernels, which will differ in having a hyperplane between the classes. The graph, predicted using Support Vector Classifier with linear kernel and RBF kernel is shown below (Figure 9.13).

9.3   CONCLUSION

Detecting acute lymphoblastic leukemia is an effective way to prevent blood cancer. In the proposed method, diagnosis of leukemia in minimum time is achieved. The automated leukocytes detecting system is developed to start treatment of affected people with the disease earlier. The preprocessing method seems to be effective with the combination of the k-means clustering technique and watershed segmentation algorithms. This is preferred to any other technique due to its simplicity and well-known clustering problem. The improved accurateness of the proposed system inspires the therapeutic field to make practice of it and this became possible by accounting for the shape features of nuclei. Feature extraction has been done, which influences the data classification. The SVM technique used is very easy to implement and ensures confidentiality of the data. The combination of these techniques provides better results. The proposed method can take advantage of all traditional techniques and achieve excellent performance.

Image processing has been an expanding field since its inception in 1994 with the ever- increasing need for medical application, speed and accuracy. Image segmentation is also part of image processing; by using clustering the severity of the cancer can be found by the shape of the cell. Diagnosis needs to be done at a very rapid rate without any compensation in loss of data or time. Hence further optimal and fast techniques can be designed for prediction in medical applications. Supervised classifications prove to be efficient but more research in the field of data mining and neural networking may increase the efficiency of prediction. It has grown in leaps and bounds with the advent of filming, entertainment and artificial intelligence, which is capable of spanning the entire globe within a few minutes. It has played key role in security purposes, agricultural sectors, video processing, digital cinema and medical applications. It could greatly aid in real-time applications with reduction in space, cost and time. This technology could be a great boon in preventing cancerous cells and also in other biomedical analysis in medicine.

REFERENCES

1. Rajendra Prasad, A. Pandey, K. P. Singh, V. P. Singh, R. K. Mishra, & D. Singh, "Retrieval of spinach crop parameters by microwave remote sensing with back propagation artificial neural networks: a comparison of different transfer functions," Advances in Space Research, vol. 50, pp. 363–370, 2012.
2. D. Simovici. (2015). "Intelligent data analysis techniques—Machine learning and data mining," In Artificial intelligent approaches in petroleum geosciences (pp. 1–51). Switzerland: Springer.
3. A. Rajaraman, & J. D. Ullman. (2012). Mining of massive datasets (vol. 77). Cambridge: Cambridge University Press.
4. J. Lin, & A. Kolcz. (2012). "Large-scale machine learning at twitter," In Proceedings of the 2012 ACM SIGMOD international conference on management of data (pp. 793–804). Scottsdale, AZ: ACM Press.
5. O. Y. Al-Jarrah, P. D. Yoo, S. Muhaidat, G. K. Karagiannidis, & K. Taha, "Efficient machine learning for big data: a review," Big Data Research, vol. 2(3), pp. 87–93, 2015.
6. Y. Tian, Y. Shi, & X. Liu, "Recent advances on support vector machines research," Technological and Economic Development of Economy, vol. U8, pp. 5–33, 2012.
7. S. Lloyd, "Least square quantization in PCM," IEEE Transaction on Information Theory, vol. 28(2), 1982.
8. Rajashree Dash, & Rasmita Dash, "Comparative analysis of K means and genetic algorithm based clustering," International Journal of Advanced Computer and Mathematical Sciences, vol. 3(2), pp. 257–265, 2012. ISSN 2230-9624. https://www.researchgate.net/publication/307863862_Comparative_Analysis_of_K-means_and_Genetic_Algorithm_based_Data_Clustering
9. B. T. Job, M. Roerdink, & A. Meijster, "The watershed transform: definition, algorithms and parallelization strategies," IOS Press Fundamental Information, vol. 41, pp. 187–228, 2001.
10. L. Zhou, "Investigate on images edge detection of pests in stored grain based on mathematical morphology," Control & Automation, vol. 4, pp. 230–5231, 2005.
11. D. Li, G. F. Zhang, Z. C. Wu, & L. N. Yi, "An edge embedded marker-based watershed algorithm for high spatial resolution remote sensing image segmentation," IEEE Transactions on Image Processing, vol. 19, pp. 2781–2787, 2010.
12. Preeti Jagadev, & H. G. Virani, "Detection of leukemia and its types using image processing and machine learning," International Conference on Trends in Electronics and Informatics, 2017. DOI: 10.1109/ICOEI.2017.8300983
13. S. Derivaux, S. Lefevre, C. Wemmert, & J. Korczak, "On Machine Learning in Watershed Segmentation," IEEE Workshop on Machine Learning for Signal Processing, 2007. DOI: 10.1109/MLSP.2007.4414304

10 Role of Machine Learning in Social Area Networks

Rajeswari Arumugam, Premalatha Balasubramaniam, and Cynthia Joseph
Coimbatore Institute of Technology

CONTENTS

10.1 INTRODUCTION TO MACHINE LEARNING

Currently, there is a high demand for artificial intelligence (AI) to create expert systems and to find solutions for complex problems in applications like recognition, natural language processing, health care, structural engineering, environmental engineering and automotive industries. Machine learning (ML) is a

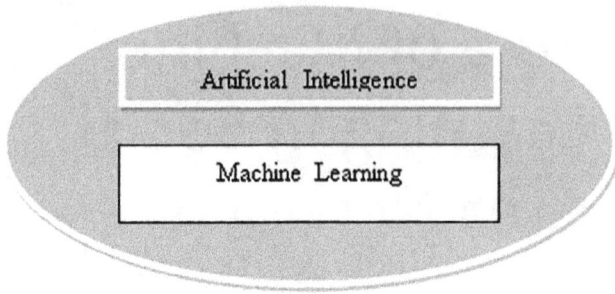

FIGURE 10.1   Relationship between artificial intelligence and machine learning.

subset of AI that allows software applications to become more precise in the prediction of outcomes without being programmed explicitly. In a nutshell, ML mimics the brain activity of human behavior and has equal computational power once it is programmed effectively. The relationship between AI and ML is shown in Figure 10.1.

ML is defined as a computer program that can adjust to new data without any intervention of human action, and it allows the machine to learn from examples and experience. ML is the logical study of algorithms and arithmetical models that the system uses to perform a particular task depending on patterns and inference. Based on training data, ML algorithms build a model to make predictions [1]. A ML algorithm is used in diverse applications where it is hard to develop an algorithm to perform the task efficiently. ML can be compared to conventional learning. In conventional learning, the input data and program are fed to the computing system where the output depends on the algorithms used. In ML, the input data and the output (trained) are fed to the computing system. The performance of ML applications not only depends on the algorithms used but also depends on various architectures. ML architectures are basically neural network architecture. A strong knowledge in linear algebra, random processes and statistics are required to implement the ML applications effectively. The difference between conventional learning and ML is shown in Figure 10.2.

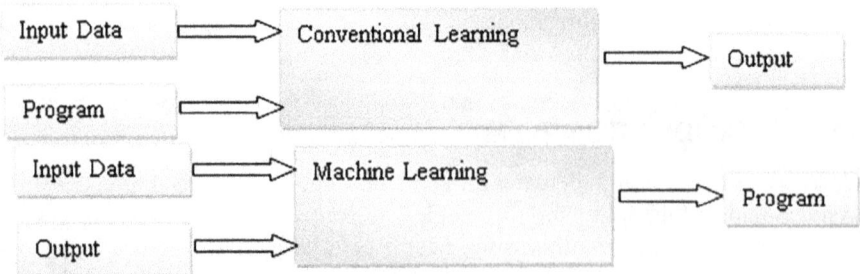

FIGURE 10.2   Difference between conventional programming and machine learning.

10.1.1 FRAMEWORK TO DEVELOP MACHINE LEARNING MODELS

The frame to develop a ML model can be divided into five stages:

- Problem identification
- Relevant data collection
- Preprocessing of data
- Building the ML model
- Deployment of model

- **Problem Identification**

A good ML project begins with the ability to define the problem clearly. In this stage the domain-specific knowledge of the expert plays a vital role in deploying the project. The major challenge in this stage is to define the right problem statement.

- **Relevant Data Collection**

The relevant data should be identified and collected once the problem statement is defined clearly. The relevant data collection is also called feature extraction. Feature extraction is defined as the process to extract the attributes from different sources that are necessary for developing the ML algorithm. The major challenge in this stage is collection data with high quality.

- **Preprocessing of Data**

In this process the noise is removed and the clarity of the data is enhanced. The collected data should be brought into suitable form for further processing.

- **Building ML Model**

The main objective of this stage is to identify a model that is suitable for the given problem. It may not always be more accurate. To find the best ML model analytical tools and solution procedures are to be strictly carried out to avoid overfitting problems. The final model for deployment is based on accuracy, speed of computation and deployment cost.

- **Deployment of Model**

Once the final model is chosen, the strategy to deploy the model has to be decided. Examples are robots, chatbots, simple action rules and so on.

10.1.2 TWO PHASES OF MACHINE LEARNING

ML consists of two important phases, the training phase and the testing phase. In the training phase the machine learns logic from the input data. More data are used in the training phase, which decides the patterns for future predictions in ML problem. The greater the amount of training data, the better will be the performance accuracy in prediction. In general, the percentage of training data from the complex datasets

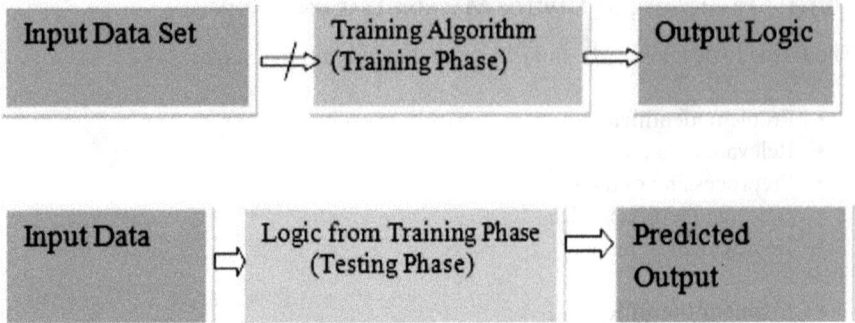

| Input Data Set | Training Algorithm (Training Phase) | Output Logic |

| Input Data | Logic from Training Phase (Testing Phase) | Predicted Output |

FIGURE 10.3 Training and testing phase in machine learning.

can be 80% or 70% or 60% and the corresponding testing samples in the proportion of 20% or 30% or 40%. The testing phase predicts the output from the learned logic for the given input. The proportion of training data and testing data should be selected in such a way that the system should produce optimized result. During the testing phase, if the predicted data has an error, the corresponding testing data should be considered as an additional training data and brought under the training set, which forms a new pattern for further predictions. A schematic representation of the training phase and testing phase are shown in Figure 10.3.

10.1.3 TYPES OF MACHINE LEARNING

The main process of ML is said to be learning from input data that is fed in. The three main types of ML algorithms are supervised learning, unsupervised learning and reinforcement learning. The three types of ML algorithms are shown in Figure 10.4.

10.1.3.1 Types of Supervised Learning

Supervised learning is defined as the task of learning a function from labeled or predetermined training data with proper guidance. Classification, linear regression, logistic regression and discriminant analysis are the types of supervised learning. These algorithms require the knowledge of both input and output variables. The prediction is achieved with the knowledge of actual values of outcome variables.

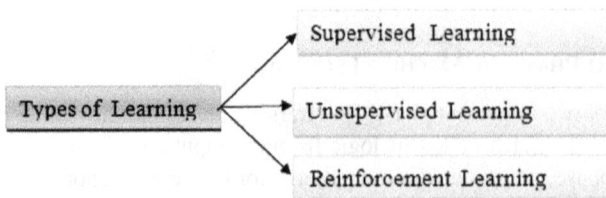

FIGURE 10.4 Types of learning algorithms.

TABLE 10.1

Example for Supervised Learning

Serial No.	Color	Shape	Plant Height in cm	Vegetable
1	Purple	Oval	45–60	Brinjal
2	Light brown	Tuber	100	Potato
3	Red	Round	182	Tomato
4	Green	Oblong	400	Lady's fingers
5	Yellow	Elliptical	304	Lemon

Table 10.1 depicts an example of a supervised learning model. Table 10.1 shows the identification of vegetables with their color, shape and plant height as features or patterns. The model has to be fed with color or shape or plant height as patterns or features. Based on the details given, the vegetable will be identified. For example,

1. **Color: Green**
2. **Shape: Oblong**
3. **Plant height: 400**
4. **The identified vegetable is: Lady's finger**

The color, shape and plant height are fed to the system as patterns for further prediction. The system, which is already trained with known patterns, tries to compare the given input with the known patterns to predict the vegetable. In this case as per the example mentioned above, the predicted vegetable is lady's finger. Similarly, all the vegetables can be identified from the features or patterns given in the Table 10.1. Colors such as purple, light brown, red, green and yellow are considered as one of the features for supervised learning. The shapes of vegetables considered for experimentation are oval, tuber, round, oblong and elliptical. Plant height for various plants lies in 45–60 cm for brinjal, 100 cm for potato, 182 cm for tomato, 400 cm for lady's fingers and 304 cm for lemon.

Figure 10.5 shows the model for supervised learning with the example of identifying brinjal from the table mentioned above.

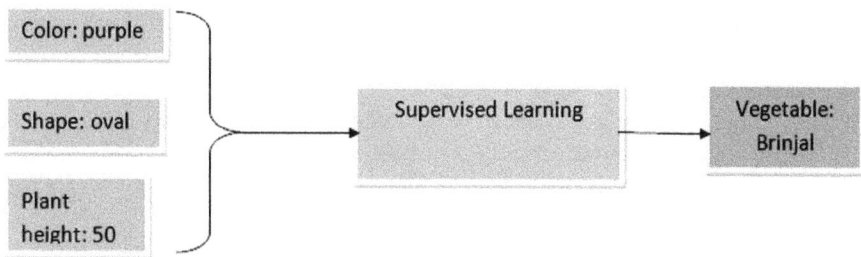

FIGURE 10.5 Model of supervised learning.

10.1.3.1.1   Classification

Classification in ML is defined as the identification of subcategories in which the new data belongs, based on the training set data patterns or features. A good example to illustrate the concept of classification in ML is a remedial class recommendation system that determines whether a student requires remedial class to improve his or her academic performance. The features or patterns considered to implement this ML are the student's test marks. The output classification categories are less than 50 marks and greater than 50 marks. Students with less than 50 marks fall under the category of remedial class. Students with greater than 50 marks fall under the category of no remedial class. An algorithm that classifies the given input in the pre-scribed category/features/patterns is called a classifier [2]. Sometimes, the classifier can be represented by a mathematical function.

10.1.3.1.2   Linear Regression

Linear regression is well-known and well-understood in the fields of statistics and ML. It is also similar to the discrimination problem in the field of engineering. The linear regression method is also known as the discriminative learning method. This deals with the optimization of input parameters using the gradient descent method and the support vector machine method. The linear regression method is used to overcome the classification problems of ML such as overfitting and the bias variance trade-off. The mathematical representation of linear regression is given by the normal straight-line equation, $y = Ax + B$, where A represents the weights and B represents the bias. The most effective way to implement a ML system is to state that the problem should be linearly separable, that is, defined by the hyperplane. For example, from the truth table of AND and OR gates, it can be defined as a linear separable problem, i.e., the hyperplane defines the class 0 and class 1 separately. In case of XOR gate, the hyperplane defines the classes of 0 and 1 in a non-linear manner, so it becomes too complicated to process the XOR gate problem in ML. To resolve this problem, optimization comes into effect and can be carried out using the gradient search method.

10.1.3.1.3   Logistic Regression

Logistic regression is also a classification algorithm that assigns patterns or features to a discrete set of classes. It is also called the binary classification method which deals with logic 0 and logic 1. The output of logistic regression should be a logic 0 or logic 1. To achieve this, the thresholding or limiting function is used. The thresholding or limiting functions are sigmoid and tangent functions. The process flow of logistic regression includes the weighted sum of the inputs and limiting function. Based on the weights used in the first level of ML architecture, the input fed to the limiting function gets varied. Before feeding the input data to the system, the mathematical function in terms of thresholding logic has to be framed properly. Then, for the set of inputs, the output should be clearly defined from the system. For experimentation the input data can be collected from the sensor devices and after proper conversion, using a suitable analog-to-digital converter, the process of learning starts. For example, take a flu detection system based on body temperature. The system detects whether a person is affected by flu or not. The patterns/features used in this system are body temperature and this system can be used for all humans irrespective of age. If the body temperature is above 97°C, it is predicted that the person has the flu, which lies under the class of logic 1. If the body temperature is below 97°C, it is predicted as the person does not have the flu, which lies under the class of logic 0.

10.1.3.2 Types of Unsupervised Learning

Unsupervised learning is defined as the training of machines without any guidance, i.e., with no predetermined classifications or labeled data. No predetermined patterns are used in unsupervised learning methods. The goal of unsupervised learning is to discern natural groupings in high dimensional data. Unlike supervised learning, there is no intended output or target to which to match the input data. This kind of learning considers only the input data. Dimensionality reduction and clustering are the two types of unsupervised learning. These algorithms do not have knowledge of the outcome variable in the dataset. They should find the possible values of outcome. Because the values of the outcome data are not known previously in training data, supervision using that knowledge is not possible. Figure 10.6 depicts an example of unsupervised learning.

10.1.3.2.1 Dimensionality Reduction

In MLa problem occurs during classification due to having to consider many factors from which final classification is performed. The factors that are considered in classification are basically variables called features. If the number of factors or features increases, the visualization and working of the training set will be more difficult. This problem is overcome by dimensionality reduction. The definition of

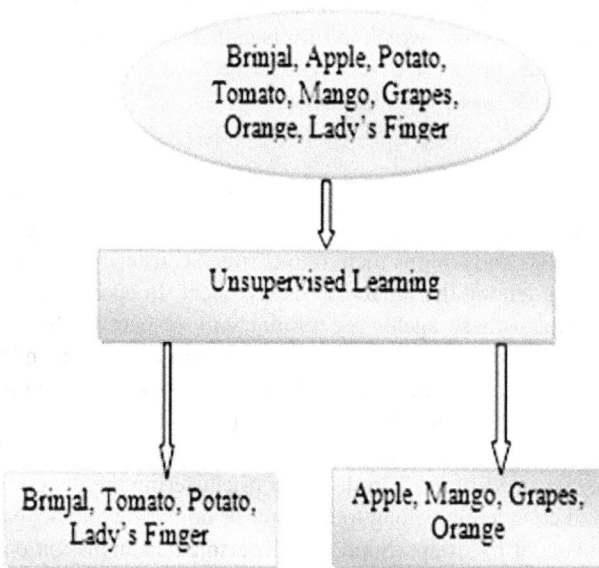

FIGURE 10.6 Model of unsupervised learning.

dimensionality reduction is to reduce the number of factors or features under consideration. The dimensionality-reduction approach is divided into feature selection and feature extraction.

10.1.3.2.2 Feature Selection

Smaller subsets are obtained from the original set of factors or features to model the problem. This is accomplished in three steps as follows: Filtering, Wrapping and Embedding.

10.1.3.2.3 Feature Extraction

The data in high dimensional space is reduced to a lower dimensional space.

10.1.3.2.4 Methods of Dimensionality Reduction

The methods used for dimensionality reduction are

- Principal Component Analysis (PCA)
- Linear Discriminant Analysis (LDA)
- Generalized Discriminant Analysis (GDA)

The importance of dimensionality reduction in ML is to reduce the unwanted features in complex data sets. It is similar to filtering noise in signal and image processing applications. The main drawback of unwanted features is that they are time consuming and require more resources with no effect in the output performance. As an example of dimensionality reduction, in signal processing applications all the inputs are multiplied with corresponding weights and summed together to produce the output. Some of the inputs have zero coefficients and do not produce any change in the output behavior. Removal of zero coefficients represents a reduction in the number of coefficients, which in turn is called dimensionality reduction. The same concept can be applied in ML to reduce the time and area complexity which decides the overall ML system performance.

10.1.3.2.5 Clustering

Clustering is grouping similar data points together. Clustering can be divided into two subcategories, hard clustering and soft clustering. Clustering segregates groups with similar qualities and assigns them into clusters. Clustering determines the fundamental grouping among the unlabeled data present. In each group, the members are called agents and these agents are similar with respect to their behavior. For example, consider a basket full of colored balls such as red, green and blue balls all mixed together. The output clusters are three in number and each cluster collects either red balls, green balls or red balls.

10.1.3.2.5.1 *Hard Clustering* In this type of clustering the data point present in the space should completely belong to a cluster or not. For example, each student is put into 1 group out of 15 groups. Suppose a department in an institution offers membership in NSS, NCC, RSP and Sports club. A very few students (e.g., 10 students) in a class completely belongs to NSS club activities, another group of 10 students in the same class completely participate in NCC activities, another group of 10 students in the class completely participate in RSP activities and the rest of the students actively participate in only Sports club activities.

10.1.3.2.5.2 Soft Clustering In this type of clustering, the data point present in the space is assigned to a cluster based on the probability or likelihood of that data point. For example, each student is assigned a probability to be in either of 15 clusters. For example, each student is put into one group out of the 15 groups in a probabilistic manner. Consider the various clubs of institution are NSS, NCC, RSP and Sports club. The probability of students from a class for cluster 1 will be 0.3, i.e., 30% from the total number of students belongs to cluster 1 or NSS Club. Similarly, the probability of students from a class for cluster 2 will be 0.2, i.e., remaining 20% of students belongs to cluster 2 or apart from cluster 1or NCC Club. Similarly, the probability of students from a class for cluster 3 will be 0.25, i.e., remaining 25% of students belongs to cluster 3 or apart from cluster 1 and cluster 2 or RSP Club. Similarly, the probability of students from a class for cluster 4 will be 0.25, i.e., the remaining 25% of students belong to cluster 4 or apart from cluster 1, cluster 2 and cluster 3 or Sports Club.

The methods used for clustering are

- K-means Clustering: In this method, the value of K must be properly chosen for perfect clustering. Some knowledge of mathematics in linear algebra is required to perform this clustering.
- Hierarchical Clustering: This method is most important in many applications.

10.1.3.3   Reinforcement Learning

Reinforcement learning is a kind of dynamic programming that trains algorithms using feedback from the system. Reinforcement learning algorithms have to take sequential actions or decisions to improve cumulative reward. The best examples of reinforcement learning are: walking practice given by the parents to the child and dictionary formation in typing text messages in all mobile phones. The newly typed words are added as a tag word or label word in the dictionary and can be utilized further. Markov chain and Markov decision processes are examples of reinforcement learning.

The difference between reinforcement learning and supervised learning are:

- In supervised learning the model is trained with the right answer because the correct answer is available in terms of patterns with the training data.
- In reinforcement learning there is no correct patterns in the training set, so the reinforcement agent is intended to learn from its previous experience.

The steps in reinforcement learning involve:

- Input
- Output
- Training
- Learning phase
- Decision phase

There are two types of reinforcement, positive reinforcement and negative reinforcement.

10.1.3.3.1 *Positive Reinforcement*

This type of reinforcement gives the positive effect on performance. It maximizes the strength and frequency of the performance. For example: walking practice given by the parents to the child.

10.1.3.3.2 *Negative Reinforcement*

This type of reinforcement gives the negative effect on performance. For example: remedial classes for weak learners.

10.1.4 CHALLENGES AND LIMITATIONS OF MACHINE LEARNING

The main challenges and limitations of ML is both the deficiency of data and dataset diversity. More mathematical models and equations are to be used to solve the NP-Hard problems with the help of effective ML techniques. Machines cannot learn if there is insufficient data, and a very diverse dataset lengthens the ML time. Machines cannot extract information when the dataset consists of zero or few variations between the data and thus leads to poor prediction. All the input data, training set patterns or features and testing data must be unique. No repeated patterns must be trained and tested. Sometimes unwanted data can be eliminated using suitable dimensionality-reduction methods. Architecture selection and algorithmic simplicity play a major role in ML to produce better performance. The major challenge in pursuing research in the field of ML is to select the suitable optimization input parameters and training patterns or features with help of best heuristic algorithm [3].

10.1.4.1 Applications of ML

Various applications of ML include

- Spam filtering
- Credit card fraud detection
- Check digit recognition
- Face recognition
- Image analysis
- Handwriting recognition
- Movie recommendation system
- Industrial Automation
- Object recognition
- Signal processing applications
- Embedded systems
- Crack detection in civil engineering
- Medical applications
- Surveillance
- Noise removal in lathe industries
- Noise removal in public

10.2 SOCIAL AREA NETWORKS

A social area network is an online platform in which people can build social relationship with each other in order to share their personal or career interests, activities, background or real-life connections. Today, there is a tremendous increase in usage of the Internet and social networks to exchange information between people in various parts of the world. Social media typically feature user-generated content and personalized profiles. People spend a great deal of time in social area networks like Twitter, Google, Yahoo and Facebook to express their opinions and feelings. People's behavior are influenced by their opinions and the study of this behavior focuses on the study of sentimental analysis. Various ML algorithms such as naïve Bayes, SVM and logistic regression are used to extract and analyze the social networking data to provide the essential results [4]. Social area networks provide the huge amounts of data (i.e., big data) with a mixture of both good and bad information.

Social media data can be used for:

- Customer service
- Public relations
- Promotions
- Behind the scenes look at organizations
- Advertising

10.2.1 RESEARCH AREAS IN SOCIAL NETWORKING

10.2.1.1 Sentimental Analysis

The process of determining user sentiments from social media is called text analytics and classifying the sentiment from comments is called sentimental analysis. Social media data consists of sentiments expressed by the users and can be classified either as a positive or negative. Comments with adjective words such as content, cheerful, pleased, glad, thrilled and joyful are called positive sentiments. Comments with adjective words such as depressed, downhearted, unhappy, low, heavy hearted, displeased, irritated and furious are called negative sentiments.

Positive Sentiment term frequency, PSTF,
= Total number of positive sentiment / Total number of sentiments

Negative Sentiment term frequency, NSTF
= Total number of negative sentiment / Total number of sentiments

10.2.1.2 Recommender System

Recommendation systems are a set of algorithms that suggest actions to the user based on the rules. It acts on behavioral data such as age, movie ID and movie genre to predict the recommendations. Rule-based algorithms (RBA) are widely used to build a recommendation system. Recommender systems usually make use of either collaborative filtering or content-based filtering or both. The other systems are knowledge-based systems. In a collaborative-filtering approach, a model can be built from a user's past behavior and as well as similar decisions made by other users. Based on the user's interest, this model can be used to predict items. The content-based filtering approach utilizes a series of discrete, pre-tagged characteristics of an item. This helps in recommending additional items with similar properties. Current recommender systems usually combine one or more approaches into a hybrid system.

10.3 INTRODUCTION TO RULE-BASED ALGORITHMS

A rule-based system is also called as a skilled system that uses rules as the knowledge representation, and these rules are coded into the system using if-then-else statements. The main idea of RBA is that knowledge is encoded as rules. It comes under the class of supervised learning. Schematic representation of RBA is shown in Figure 10.7.

10.3.1 EXPERIMENT 1: SENTIMENTAL ANALYSIS

In this experiment three emotions are categorized as sad, happy and angry with respect to the adjective words. For sentimental analysis on social media adjective words are considered to frame the rule for classification. Binary codes are assigned to the adjective words for experimentation. To incorporate the real-time scenario binary codes are randomly generated and fed as an input to the RBA [5].

Emotion input, x = round (rand (1, k); where k = 5 is used in this chapter (1)

Classification is done by using switch and case statement in this experiment. Adjective words for emotion classification and the binary codes considered for experimentation are shown in Table 10.2. The following parameters are considered as input for experimentation: three classes such as sad, happy and angry are considered for emotion classification and the adjectives are assigned based on the three classes.

For example, the adjective words for class 1 (i.e., sad) are depressed, downhearted, unhappy, dull, heavy hearted, low, sorrowful, cheerless, joyless, regrettable, mournful, shameful and bad. Similarly, the adjective words for class 2 (i.e., happy) are content, cheerful, pleased, glad, thrilled, elated, joyful and bliss. The adjective words for class 3 (i.e., angry) are irritated, furious, displaced, ferocious, bitter, annoyed and impassioned.

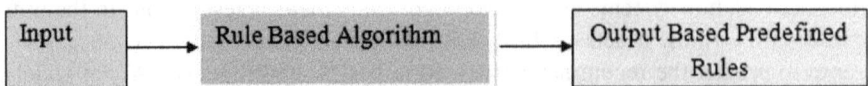

FIGURE 10.7   Schematic representation of rule-based algorithms.

TABLE 10.2

Adjective Words for Emotion Classification and Their Binary Codes

Class 1	Binary Code	Class 2	Binary Code	Class 3	Binary Code
Sad (Negative Sentiment)		Happy (Positive Sentiment)		Angry (Negative Sentiment)	
Depressed	00001	Content	01110	Irritated	10110
Downhearted	00010	Cheerful	01111	Furious	10111
Unhappy	00011	Pleased	10000	Displeased	11000
Dull	00100	Glad	10001	Ferocious	11001
Heavy hearted	00101	Thrilled	10010	Bitter	11010
Low	00110	Elated	10011	Annoyed	11011
Sorrowful	00111	Joyful	10100	Impassioned	11100
Cheerless	01000	Bliss	10101		
Joyless	01001				
Regrettable	01010				
Mournful	01011				
Shameful	01100				
Bad	01101				

The proposed model for sentimental analysis is shown in Figure 10.8. The model consists of a binary pattern as input where the inputs are given in the binary format, for example: 01000. The given input is fed to the RBA where the emotions are predicted with the help of the adjective words assigned. From the above-mentioned binary pattern 01000, the adjective word assigned is cheerless that belongs to the sad class. So the predicted output will be displayed as sad. Similarly, all the emotions can be identified with the help of their respective adjective words.

10.3.1.1 Sample Experimentation Results

Experiments have been carried out for the emotions like angry, sad and happy.

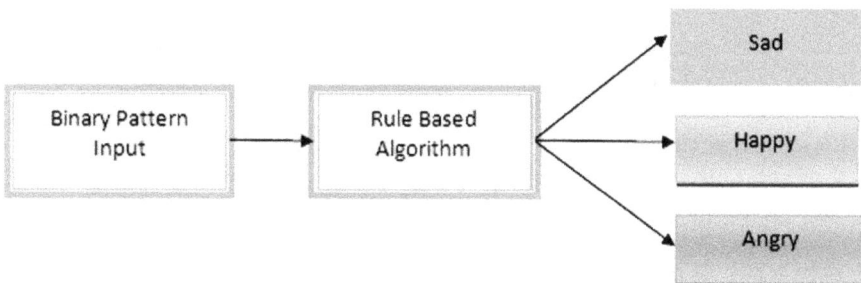

FIGURE 10.8 Proposed model for sentimental analysis in social area networking.

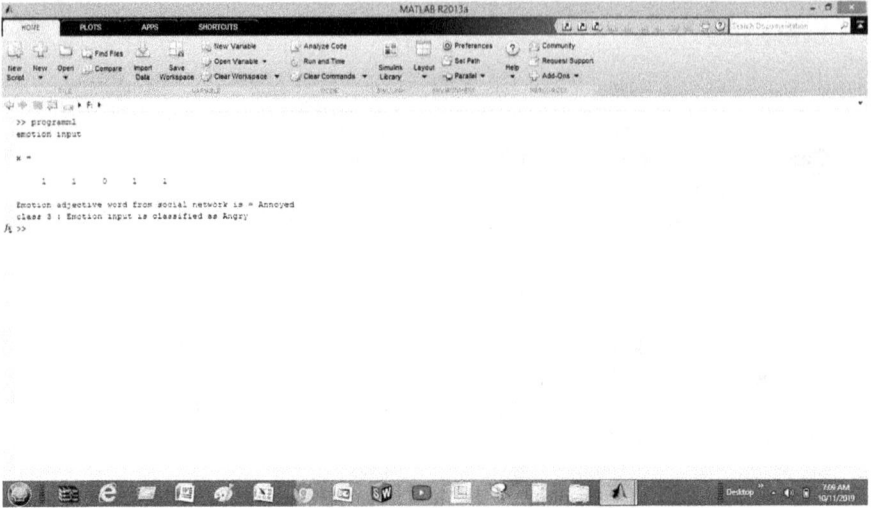

FIGURE 10.9 Simulation result for the binary pattern 11011.

Figure 10.9 shows that the binary pattern 11011 is fed as an input and the adjective word assigned is annoyed that belongs to the angry class. So, the output is displayed as angry.

Figure 10.10 shows that the binary pattern 01101 is fed as an input and the adjective word assigned is bad that belongs to the sad class. So, the output is displayed as sad.

Figure 10.11 shows that the binary pattern 00111 is fed as an input and the adjective word assigned is sorrowful that belongs to the sad class. So, the output is displayed as sad.

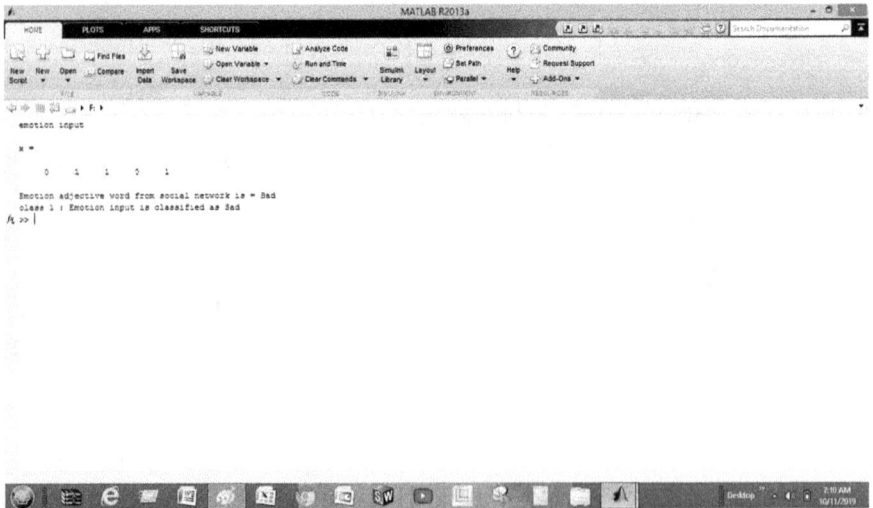

FIGURE 10.10 Simulation result for the binary pattern 01101.

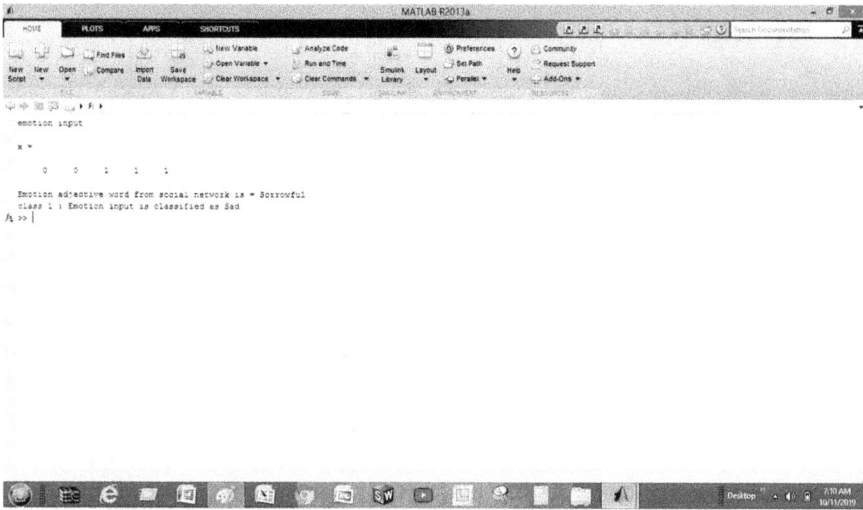

FIGURE 10.11 Simulation result for the binary pattern 00111.

Figure 10.12 shows that the binary pattern 11010 is fed as an input and the adjective word assigned is bitter that belongs to the angry class. So, the output is displayed as angry.

Figure 10.13 shows that the binary pattern 10000 is fed as an input and the adjective word assigned is pleased that belongs to the happy class. So, the output is displayed as happy.

Figure 10.14 shows that the binary pattern 00110 is fed as an input and the adjective word assigned is low that belongs to the sad class. So, the output is displayed as sad.

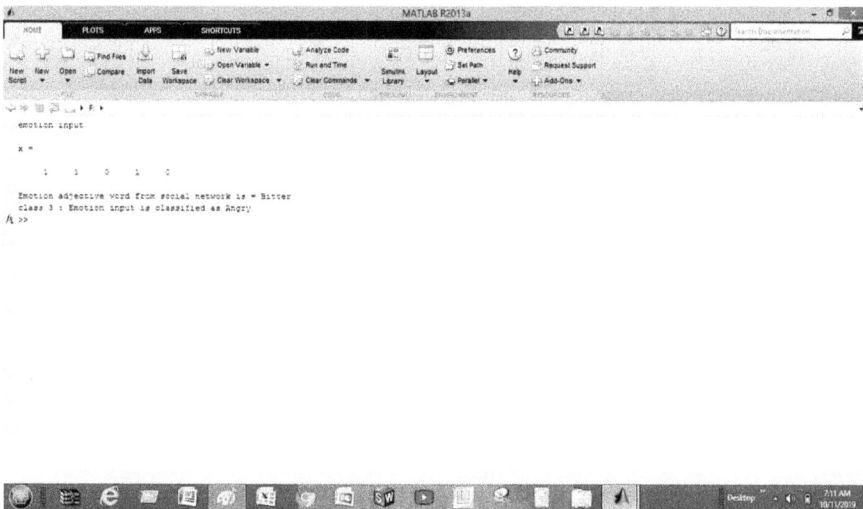

FIGURE 10.12 Simulation result for the binary pattern 11010.

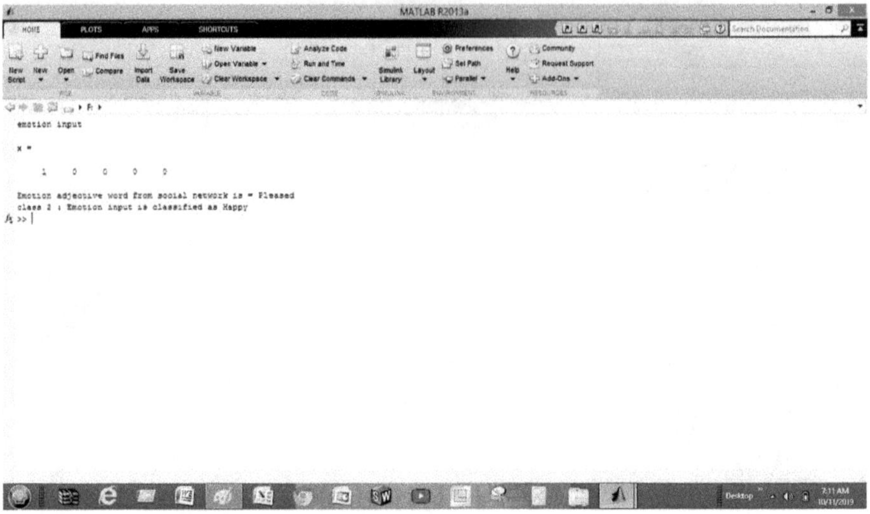

FIGURE 10.13 Simulation result for the binary pattern 10000.

Figure 10.15 shows that the binary pattern 01100 is fed as an input and the adjective word assigned is shameful that belongs to the sad class. So, the output is displayed as sad.

Figure 10.16 shows that the binary pattern 11000 is fed as an input and the adjective word assigned is displeased that belongs to the angry class. So, the output is displayed as angry.

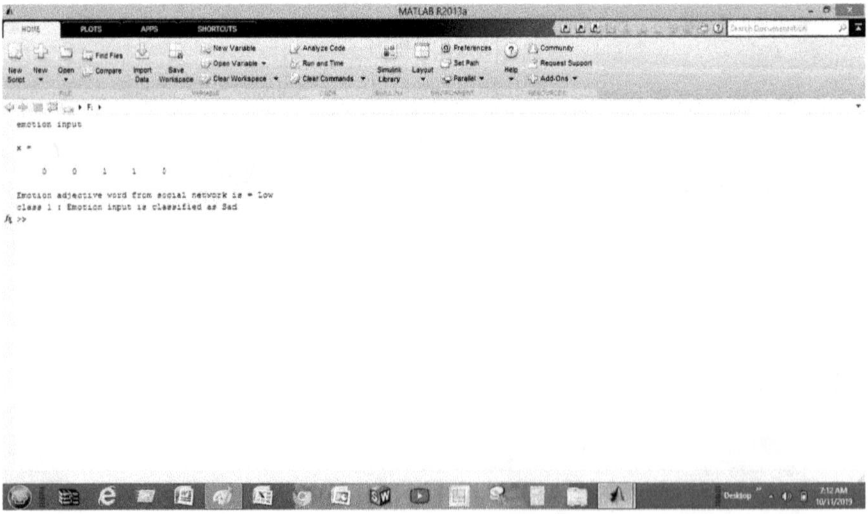

FIGURE 10.14 Simulation result for the binary pattern 00110.

FIGURE 10.15 Simulation result for the binary pattern 01100.

Figure 10.17 shows that the binary pattern 10101 is fed as an input and the adjective word assigned is bliss that belongs to the happy class. So, the output is displayed as happy.

The outputs shown in Figures 10.9, 10.10, 10.11, 10.12, 10.13, 10.14, 10.15, 10.16 and 10.17 are tabulated in Table 10.3.

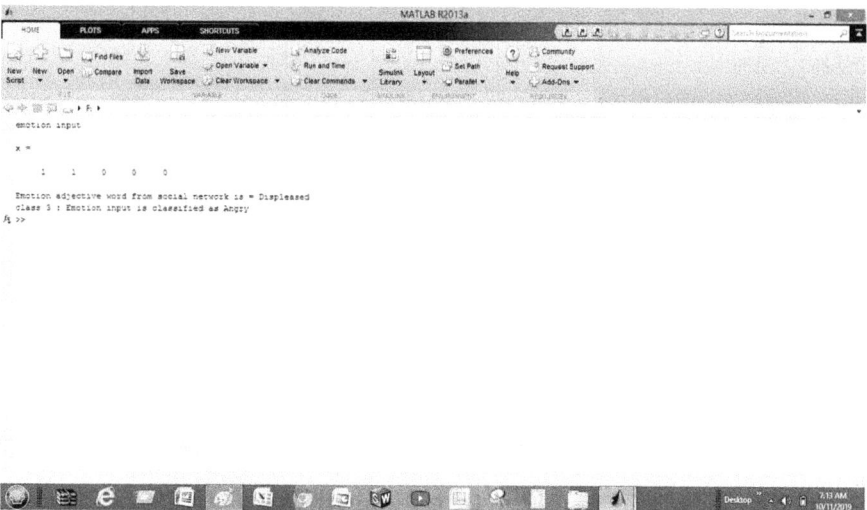

FIGURE 10.16 Simulation result for the binary pattern 11000.

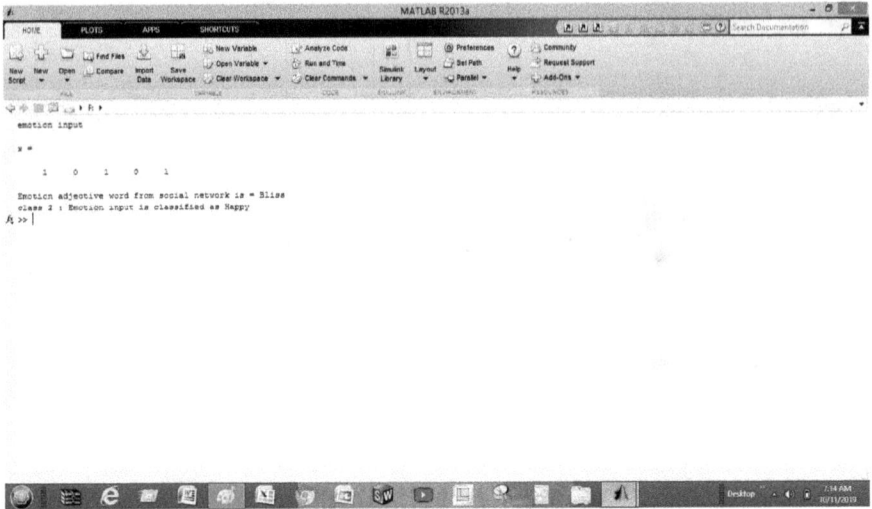

FIGURE 10.17 Simulation result for the binary pattern 10101.

10.3.2 EXPERIMENT 2: MOVIE RECOMMENDATION

In this chapter, a movie recommendation system has been implemented with respect to genre of movie such as adventure, comedy, romance, comedy and romance and adventure and romance. For recommendation analysis on social media, the age of the user and movie ID are considered to frame the rule for classification.

Classification is done by using if-then-else statements. Movie ID, age of the user and genre of the movie are considered for experimentation as shown in Table 10.4. The proposed model for movie recommendation analysis is shown in Figure 10.18. Movie ID and age are given as the input to the model. The inputs are given to the RBA that suggests whether movie is recommended or not recommended. For example, if the

TABLE 10.3
Summary of Sentimental Analysis

Serial No.	Emotion Input	Adjective Word	Class	Emotion Output
1	11011	Annoyed	3	Angry
2	01101	Bad	1	Sad
3	00111	Sorrowful	1	Sad
4	11010	Bitter	3	Angry
5	10000	Pleased	2	Happy
6	00110	Low	1	Sad
7	01100	Shameful	1	Sad
8	11000	Displeased	3	Angry
9	10101	Bliss	2	Happy

TABLE 10.4

Experimentation Details for Movie Recommendation System

Movie ID	Genre of Movie	Age Limit	Recommendation/ Not Recommended
1.	Adventure	13–60	Recommended
2.	Comedy	05–70	Recommended
3.	Romance	18–70	Recommended
4.	Comedy and romance	18–70	Recommended
5.	Adventure and romance	18–70	Recommended

movie ID is 1 and the age is 45 then the genre of the movie is displayed as adventure. Since the age is in the range of 13–60 it displays recommended.

10.3.2.1 Sample Experimentation Results

Experimentation has been carried out for the recommendation of movie.

In Figure 10.19 the movie ID and age are entered as 5 and 17. The genre of the movie is displayed as adventure and romance and the movie is not recommended.

In Figure 10.20 the movie ID and age are entered as 5 and 85. The genre of the movie is displayed as adventure and romance and the movie is not recommended.

In Figure 10.21 the movie ID and age are entered as 5 and 25. The genre of the movie is displayed as adventure and romance and the movie is recommended.

In Figure 10.22 the movie ID and age are entered as 4 and 10. The genre of the movie is displayed as comedy and romance and the movie is not recommended.

In Figure 10.23 the movie ID and age are entered as 4 and 65. The genre of the movie is displayed as comedy and romance and the movie is recommended.

In Figure 10.24 the movie ID and age are entered as 3 and 10. The genre of the movie is displayed as romance and the movie is not recommended.

In Figure 10.25 the movie ID and age are entered as 4 and 55. The genre of the movie is displayed as comedy and romance and the movie is recommended.

In Figure 10.26 the movie ID and age are entered as 3 and 2. The genre of the movie is displayed as romance and the movie is not recommended.

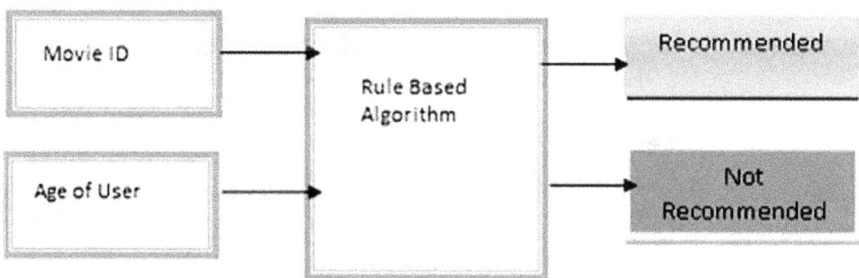

FIGURE 10.18 Proposed model for movie recommendation.

```
enter your movie id: '5'
enter your age: 17
 Genre   of movie is is Adventure and Romance
Movie is not recommended to you>>
```

FIGURE 10.19   Simulation result for the input of movie ID and age as 5 and 17.

```
enter your movie id: '5'
enter your age: 85
Genre    of movie is is Adventure and Romance
Movie is not recommended to you>>
```

FIGURE 10.20   Simulation result for the input of movie ID and age as 5 and 85.

```
enter your movie id: '5'
enter your age: 25
Genre    of movie is is Adventure and Romance
Movie is recommended to you>> |
```

FIGURE 10.21   Simulation result for the input of movie ID and age as 5 and 25.

```
enter your movie id: '4'
enter your age: 10
 Genre of movie id is Comedy and Romance
Movie is not recommended to you>> |
```

FIGURE 10.22   Simulation result for the input of movie ID and age as 4 and 10.

```
enter your movie id: '4'
enter your age: 65
 Genre of movie id is Comedy and Romance
Movie is recommended to you>>
```

FIGURE 10.23   Simulation result for the input of movie ID and age as 4 and 65.

```
enter your movie id: '3'
enter your age: 10
 Genre  of movie id is Romance
Movie is not recommended to you>>
```

FIGURE 10.24   Simulation result for the input of movie ID and age as 3 and 10.

```
enter your movie id: '4'
enter your age: 55
Genre  of movie id is Comedy and Romance
Movie is recommended to you>>
```

FIGURE 10.25   Simulation result for the input of movie ID and age as 4 and 55.

```
enter your movie id: '3'
enter your age: 2
Genre  of movie id is Romance
Movie is not recommended to you>>
```

FIGURE 10.26   Simulation result for the input of movie ID and age as 3 and 2.

```
enter your movie id: '3'
enter your age: 45
 Genre  of movie id is Romance
Movie is recommended to you>>
```

FIGURE 10.27   Simulation result for the input of movie ID and age as 3 and 45.

```
enter your movie id: '2'
enter your age: 5
 Genre  of movie id is Comedy
Movie is recommended to you>> |
```

FIGURE 10.28   Simulation result for the input of movie ID and age as 2 and 5.

In Figure 10.27 the movie ID and age are entered as 3 and 45. The genre of the movie is displayed as romance and the movie is recommended.

In Figure 10.28 the movie ID and age are entered as 2 and 5. The genre of the movie is displayed as comedy and the movie is recommended.

In Figure 10.29 the movie ID and age are entered as 2 and 80. The genre of the movie is displayed as comedy and the movie is not recommended.

```
enter your movie id: '2'
enter your age: 80
 Genre  of movie id is Comedy
Movie is not recommended to you>>
```

FIGURE 10.29   Simulation result for the input of movie ID and age as 2 and 80.

```
enter your movie id: '1'
enter your age: 11
 Genre of movie id is Adventure
Movie is not recommended to you>>
```

FIGURE 10.30   Simulation result for the input of movie ID and age as 1 and 11.

```
enter your movie id: '1'
enter your age: 48
 Genre  of movie id is Adventure
Movie is recommended to you>> |
```

FIGURE 10.31   Simulation result for the input of movie ID and age as 1 and 48.

```
enter your movie id: '1'
enter your age: 60
Genre  of movie id is Adventure
Movie is recommended to you>>
```

FIGURE 10.32   Simulation result for the input movie ID and age as 1 and 60.

In Figure 10.30 the movie ID and age are entered as 1 and 11. The genre of the movie is displayed as adventure and the movie is not recommended.

In Figure 10.31 the movie ID and age are entered as 1 and 48. The genre of the movie is displayed as adventure and the movie is recommended.

In Figure 10.32 the movie ID and age are entered as 1 and 60. The genre of the movie is displayed as adventure and the movie is recommended.

The outputs shown above are tabulated in Table 10.5

TABLE 10.5
Summary of Movie Recommendation

Input	Movie ID Entered	Age of User	Genre of Movie Identified	Output
Sample 1	5	17	Adventure and Romance	Not Recommended
Sample 2	5	85	Adventure and Romance	Not Recommended
Sample 3	5	25	Adventure and Romance	Recommended
Sample 4	4	10	Comedy and Romance	Not Recommended
Sample 5	4	65	Comedy and Romance	Recommended
Sample 6	3	10	Romance	Not Recommended
Sample 7	4	55	Comedy and Romance	Recommended
Sample 8	3	2	Romance	Not Recommended
Sample 9	3	45	Romance	Recommended
Sample 10	2	5	Comedy	Recommended
Sample 11	2	80	Comedy	Not Recommended
Sample 12	1	11	Adventure	Not Recommended
Sample 13	1	48	Adventure	Recommended
Sample 14	1	60	Adventure	Recommended

10.4 CONCLUSION AND FUTURE SCOPE

ML plays a major role in social area networking to predict the future decisions in handling big data. In this chapter, two main issues in social networking such as sentimental analysis and movie recommendations are implemented with the help of RBA. The future scope of this chapter is to implement high-level ML algorithms to solve the marketing and promotion issues in social area networking.

REFERENCES

1. Sumit Das, Aritra Dey, Akash Pal, Nabamita Roy (2015), "Applications of Artificial Intelligence in Machine Learning: Review and Prospect", International Journal of Computer Applications (0975–8887), vol. 115, no. 9, pp. 31–41.
2. O. Simeone (2018), "A Very Brief Introduction to Machine Learning with Applications to Communication Systems", IEEE Transactions on Cognitive Communications and Networking, vol. 4, no. 4, pp. 648–664, doi: 10.1109/TCCN.2018.2881442.
3. Pinky Sodhi, Naman Awasthi, Vishal Sharma (2019), "Introduction to Machine Learning and Its Basic Application in Python", Proceedings of 10th International Conference on Digital Strategies for Organizational Success, doi: 10.2139/ssrn.3323796.
4. Suha AlAwadhi, Peter Parycek, Jay P. Kesan (2013), "Introduction to Social Media and Social Networking Minitrack", IEEE Explore, 46th Hawaii International Conference on System Sciences, doi: 10.1109/HICSS.2013.325.
5. K. Ahmed, N. E. Tazi, A. H. Hossny (2015), "Sentiment Analysis over Social Networks: An Overview", IEEE International Conference on Systems, Man, and Cybernetics, Kowloon, pp. 2174–2179. doi: 10.1109/SMC.2015.380.

11 Breast Cancer and Machine Learning

Interactive Breast Cancer Prediction Using Naive Bayes Algorithm

Atapaka Thrilok Gayathri and
Samuel Theodore Deepa
Shri Shankarlal Sundarbai Shasun Jain College for Women

CONTENTS

11.1 INTRODUCTION TO BREAST CANCER

Cancer is all about the changes that happen when human body cells develop abnormally. The human body is the composed of minute blocks called cells. These cells usually build in the body whenever necessary and get expired when they are not necessary. Cancer cells are unnatural cells that get build in the human body. Commonly, in all cancer types these unnatural cells grow unlimitedly and form a lump in the human body called a tumor.

Often the abnormal cells are formed in either lobules or ducts of the breast. Lobules are the glands in the women that produce milk whereas ducts are of the channels that bring milk to the nipple. Fatty tissue or fibrous connective tissues are other places cancer can occur. These uncontrolled cancer cells even travel to the lymph nodes under the arms, from these lymph nodes cancer cells move to other parts of the body. When breast cancer spreads or breast cancer cells move to other parts of the body through the blood vessels or lymph vessel then it is called as metastasis.

11.1.1 Types of Breast Cancer

The classifications of breast cancer are of two types: invasive or noninvasive. The invasive type of cancer transfers to nearby tissues [1]. Noninvasive breast cancers do not transfer away from the milk ducts inside the breast. Cancer that steps in first to the ducts or lobes is termed as ductal carcinoma or lobular carcinoma, respectively.

- **Ductal carcinoma.** Majority or most part of breast cancer of this type appear in the cells lining the milk ducts.
  - **Ductal carcinoma in situ (DCIS).** A kind of disease usually present alone in the duct.
  - **Invasive or infiltrating ductal carcinoma.** A kind of disease explicitly transferred beyond the duct.
- **Lobular carcinoma.** This kind of disease or problem which appears first within the lobules.
  - **Lobular carcinoma in situ (LCIS).** The condition when abnormal cells are present only in the lobules is called LCIS, which is not considered as cancer. However, there is the problem of stimulating invasive breast cancer in the breast by this LCIS.
  - **Invasive lobular carcinoma.** When lobular carcinoma expands beyond the duct.

11.1.1.1 Breast Cancer Symptoms

There are several symptoms for breast cancer; one of the most important is a lump or thick breast tissue.

Some of the common symptoms are as follows:

- Different feel in the surrounding tissue due to breast lump or thickening.
- Change in the appearance of a breast as such as size or shape.
- Dimpling on the breast.
- Modification occurring in the nipple.
- Peeling, scaling, crusting, etc. occurring on the skin around the nipple.
- Other kinds of disease or a change in color of the skin on the breast to red.

Factors that in increase the risk of breast cancer:

- Older age
- A family history
- A previous records of breast cancer
- Overweight or obesity
- Excess intake of alcohol

11.1.1.2 Treating Breast Cancer

Cancers that are diagnosed at an early stage can be treated.

Breast cancer treatment includes the combination of:

- Surgery
- Chemotherapy
- Radiotherapy

Surgery is the first type of treatment performed, followed by chemotherapy or radiotherapy or, in some cases, hormone or biological treatments [1].

11.2 INTRODUCTION TO MACHINE LEARNING ALGORITHM

One of the subfields in artificial intelligence (AI) is machine learning. The main goal
of machine learning is to understand the nature or structure of data and implement
the data in the models for analysis to determine the facts for best utility.

Even though machine learning comes under the field of computer science, it is
different from traditional systems. In traditional learning, algorithms are set with
predefined programs to analyze the data to find the solution. Whereas in machine
learning, the system is trained with a set of inputs and special algorithms are implemented to find a set of outputs. Machine learning is an automated process of decision
making in various fields (Fig.11.1).

In the field of AI the term machine learning is coined as "It gives computers the
ability to learn without being explicitly programmed."

In the field of data analytics, machine learning is used to develop complex models
and algorithms used to predict hidden facts by the researcher, data scientists. This
process is called predictive analysis. Historical relationships and trends in data are
taken into consideration for analysis.

11.2.1 CLASSIFICATION OF MACHINE LEARNING

Depending on the nature of learning machine learning is classified as:

- Supervised
- Unsupervised
- Reinforcement
- Semi-supervised

Supervised learning: Learning from the sample data and associated target responses
and trying to make predictions for new examples comes under the category of supervised learning. Sample data can be numeric values or string labels.

Unsupervised learning: Learning without sample data and associated target
responses and trying to determine the patterns is termed as unsupervised learning.
This type tries to analyze the data to determine new features that can be used as
input to the supervised learning.

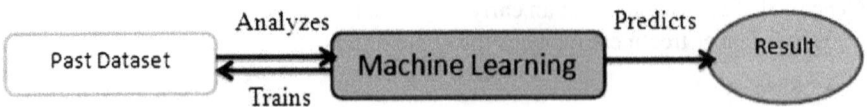

FIGURE 11.1   Process of machine learning.

Reinforcement learning: This type uses the concepts of trial and error. The algorithm is accompanied with the examples that pose both positive and negative feedback based on the solution.

Semi-supervised learning: In this type incomplete training data with missing output is given as sample data [2].

11.2.2 Grouping Machine Learning Based on the Results Obtained

Machine learning systems can also be grouped based on the result obtained after the analysis.

1. **Classification:** This type is usually carried out in a supervised way; the model that is developed by the software programmer learns from the data that is fed into the model. Based on this learning process the model creates the classification for the newly observed data.
2. **Regression:** This type is a kind of supervised problem. In this case, where the outputs are continuous rather than discrete.
3. **Clustering:** This is a type of unsupervised learning approach among machine learning algorithms where the set of input is classified into groups, where the groups are unrevealed earlier.

11.2.3 Implementing Machine Learning

Prediction: Machine learning is used as a prediction system. Machine learning is used in the field of healthcare to predict disease, for example types of breast cancer.

Image recognition: Machine learning can be used to recognize an image. For example, various X-rays needed to have their images recognized to diagnose the problem.

Speech Recognition: At the present time voice recognition has become an important aspect for the purpose of security, searching etc.

Financial industry and trading: Machine learning is a fast-moving technology used in various companies for fraud investigations and credit card security purpose.

11.3 RISK FACTORS OF BREAST CANCER

In today's world one of the most common cancers among women is breast cancer. This kind of disease usually occurs in humans when abnormal cells grow abundantly, accumulate at a fast rate, which in turn develops lumps in the human body. The travel of the abnormal cells start from the breast to the lymph node then later to the remaining parts of the body. This travel is considered to be metastasize spread. The reasons for this kind disease could be the daily lifestyle, food habits, atmosphere change factors, etc.

11.3.1 Risk Factors

Factors that can be considered to be the reason for high chances of breast cancer can be:

- **Feminine.** According to research, women are more commonly affected when compared to men.
- **Age Factor.** Increasing lifespan of an individual is one factor.

- **A previous medical history of breast ailments.** Breast conditions such as LCIS or atypical hyperplasia of the breast, can be another reason.
- **A chance of occurrence in another breast.** Breast cancer is the kind of disease that spreads, so there is a chance of occurrence in the other side, too.
- **Earlier generations of family members who had breast cancer.** Hereditary is considered to be one of the factors for breast cancer.
- **Inheritance of genes from family members considered as risk.** Genes that are inherited from parent to child are another risk factor. The most commonly known genes that are inherited are BRCA1 and BRCA2. These genes increased chances of breast cancer and other cancers.
- **Radiation treatment.** If any such kind of treatments to the chest were undergone at a young age, they can be considered as a risk factor.
- **Diet factor.** Obesity increases the chances of disease.
- **Childhood onset of menstruation.** If a woman began her periods before the age of 12, she has greater chances of breast cancer.
- **Menopause.** Menopause increases the chances of breast cancer.
- **Giving birth to a child at an older age.** When a woman delivers a child after the age of 30 she is more at risk for breast cancer.
- **Less chances of pregnancy.** Women who have less chances of pregnancy have a greater chances of breast cancer.
- **Postmenopausal hormone therapy treatment.** A woman who is under the treatment of hormone therapy with the combination of estrogens and progesterone to overcome the problems of periods increases her risk of developing breast cancer.
- **Alcohol intake.** Alcohol intake increases the chances of breast cancer [3].

11.4 TNM STAGING SYSTEM

The most common method the healthcare professionals use to elaborate the condition of breast cancer is the tumor, nodes, metastasis (TNM) system. Specialists elaborate on the outcomes obtained as a result of various examinations made, to analyze the following factors:

- **Tumor (T):** The T represents size of the primary tumor and the place where it is located.
- **Node (N):** N represents to what extend tumor moved to the lymph nodes.
- **Metastasis (M):** M represents to what extent the cancer moved to other parts of the body.

The TNM analysis explains to the cancer patient their condition and how far they are affected. There are five stages: Stage 0 (zero), which is noninvasive DCIS, and stages I through IV (1 through 4), which are used for invasive breast cancer. This procedure of analyses helps the specialist to establish proper planning for the best treatment of the patient.

11.4.1 T CATEGORY

The T category (T0, Tis, T1, T2, T3, or T4) represents the size of the tumor cells and determines whether it has been extended to the skin on the breast or whether to the chest wall below the breast. Higher T numbers indicate larger tumor cells such that large spread has occurred to the tissues.

TX: Fundamental tumor cannot be diagnosed.

T0: (T plus zero): There is no symptom of cancer cells.

Tis: Cancer is present inside the ducts or lobules in the breast without beginning its spread to remaining parts of the breast. This case is termed *carcinoma in situ*.

T1: Measurement of the tumor size is accounted to be 20 millimeters (mm) or smaller in size. This measurement is considered to be less than an inch. This category is further categorized into four different stages based on the measurement of tumor size:

- T1mi is stage where the measurement is 1 mm or smaller.
- T1a is stage where the measurement is a maximum 5 mm but minimum 1 mm or minor.
- T1b is a tumor size that is a maximum 10 mm but minimum 5 mm or minor.
- T1c is a tumor size that is a maximum 20 mm but minimum 10 mm or minor.

T2: In this case tumor size is a minimum 20 mm but not more than 50 mm.

T3: In this case tumor size with a maximum 50 mm.

T4: Additional subdivision of this stage is to be as follows.

- T4a implies the tumor cell has started its origin in the chest wall.
- T4b implies the tumor cell has started its origin in the skin.
- T4c implies the tumor cell has started its origin in both the chest wall and the skin.
- T4d is kind of group termed inflammatory breast cancer.

11.4.2 N CATEGORY

The N category (N0, N1, N2, or N3) which implies whether the cancer started its origin in the lymph nodes in the breast and, if that is the case, how many lymph nodes are affected. The higher the number the more lymph node are affected. If none of the lymph nodes were identified, then the stage in the N category is NX.

NX: None of the lymph nodes are infected.

N0: Cases falls into either of two categories:

- No such disease present in the lymph nodes.
- Measurement of cell size is smaller than 0.2 mm.

N1: The breast infection has continued to one to three auxiliary lymph nodes and/or else to the internal mammary lymph nodes.

N2: Cells are extended to four to nine auxiliary lymph nodes or cells have extended to the internal mammary lymph nodes.

N3: The infection has extended to ten or more auxiliary lymph nodes or the infection had its origin in the lymph nodes positioned beneath the clavicle, or collarbone. Or it may have its origin in the internal mammary lymph nodes. If this is the case, where the cancer has moved to the lymph nodes on top of the clavicle, is termed as supraclavicular lymph nodes, is also characterized as N3.

11.4.3 M CATEGORY

The M category in the TNM staging system describes how far the cancer has moved to remaining parts of the body, and hence is called distant metastasis.

MX: Distant spread of the cancer cannot be evaluated.

M0: Cancer has not spread yet.

M0 (i+): None of the clinical records like radiographic shows the infection as distant metastases. Microscopic records show that tumor cells are found in the blood, bone marrow, or other lymph nodes whose measurement is not more than 0.2 mm.

M1: Clinical confirmation of metastasis to remaining part of the body means that abnormal cells developed or spread in other organs of the body [4].

11.5   CANCER STAGING, GROUPING, AND GRADING

The stage of the breast cancer patient is determined by the doctor based on the combination of the T, N, and M staging system along with the tumor size grading system in addition to the output of ER/PR and HER2 test. The present knowledge about the patient condition is used by health care professionals to determine diagnosis. The simpler and the most effective approach used in evaluating the condition of the patient is by using the T, N, and M staging. This analyses method is deployed in the research work.

The condition of the patient needs to be analyzed by the doctor to save the patient; this is done after the surgery, usually after the time span of five to seven days. Neoadjuvant therapy is the treatment given to the whole body before surgery. Healthcare professionals state that Stage I to Stage IIA cancer is a starting stage, and Stage IIB to Stage III is a bit advanced.

Stage 0: When the epidemic is present only in the ducts and lobules in and around the breast tissue but has not escalated to the neighboring tissues then it is considered as Stage zero (0) or noninvasive cancer(Tis,N0,M0).

Stage IA: The tumor grading is small, invasive, but it doesn't starts its origin in lymph nodes (T1, N0, M0) is considered as Stage IA.

Stage IB: In this category sickness has been transferred to lymph nodes; further, the size of the lymph node is greater than 0.2 mm and lower than 2 mm in size. Absence of the symptoms of tumor cells in the human breast or the size of tumor cells is 20 mm or lower (T0 or T1, N1, M0).

Stage IIA: This category can fall under any of the following conditions:

- Signs or symptoms of tumor cells are absent in the breast, but the abnormal cells have expanded to 1 to 3 auxiliary lymph nodes, but the cancer has not yet moved to other organs of the body. (T0, N1, M0).

- The tumor is of size 20 mm or minor and the cancer has transferred to the auxiliary lymph nodes (T1, N1, M0).
- The tumor of size larger than 20 mm yet not beyond 50 mm but the cancer has not extended to the auxiliary lymph nodes (T2, N0, M0).

Stage IIB: This level can fall on one of these two conditions:

- The tumor is of size larger than 20 mm yet not beyond 50 mm but the cancer has extended to one to three auxiliary lymph nodes (T2, N1, M0).
- The tumor is of size beyond 50 mm, yet it has not started its origin in auxiliary lymph nodes (T3, N0, M0).

Stage IIIA: In this level cancer started its origin in four to nine auxiliary lymph nodes or in internal mammary lymph nodes, but it can be of any size; the point to be noted is it has not moved to other parts of the body (T0, T1, T2 or T3, N2, M0). In Stage IIIA tumor cells can be of size greater than 50 mm which has started its growth in one to three auxiliary lymph nodes (T3, N1, M0).

Stage IIIB: When the disease is diagnosed as inflammatory breast cancer. If this is the case, then the tumor has started to move to the chest wall with some symptoms like swelling or ulceration of the breast. There is a chance that it may not have been spread up to nine auxiliary or internal mammary lymph nodes, but it has started its origin in other organs.(T4; N0, N1 or N2; M0).

Stage IIIC: This is the condition where the cancer rate has increased, in which measurement of tumor cells constitutes any size and its growth has started in ten or more auxiliary lymph nodes, and also to the internal mammary lymph nodes, and further to the lymph nodes beneath the collarbone. Its origin has not started its growth in remaining areas of the body (any T, N3, M0).

Stage IV (metastatic): This category is the stage where the tumor cells constitute any size but it has started to move to other parts of the body like the bones, lungs, brain, etc. When this stage is diagnosed with no prior diagnosis(found to be 5% to 6%) it is termed de novo metastatic breast cancer.

Recurrent: The cancer that reappears even after treatment, which can be either local, regional, or distant. There are possibilities where the cancer may not reappear; if that is the case then further tests are conducted to determine the chances of reappearance. Tests or scans are carried out similarly over the period of diagnosis.

11.6 STAGING AND GRADING

Staging and grading is the methodology used by doctors to explain to the patient how far the patient is affected with the breast cancer and how far the cancer grows. The staging and grading process helps the healthcare professional team plan the best treatment for the patient. This methodology is useful even after surgery. Grading is the methodology that indicates what the breast cancer cells look like. It is categorized in to 1, 2, 3 categories (Fig.11.2).

Grade 1 – In this classification type the cancer cells looks like normal cells, and the growth is generally moderate compared to the other categories.

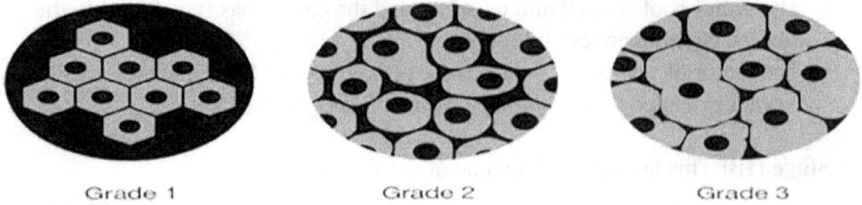

FIGURE 11.2   Types of grade cells.

Grade 2 – In this classification type the cancer cells are moderately larger than natural cells, with different pattern and growth is rapid compared to normal cells.

Grade 3 – In this classification type the cancer cells look different from normal cells, and growth is usually faster.

Grading methodology for noninvasive breast cancers like DCIS is completely different, with the categories low, medium, or high.

11.6.1   STAGING

Staging methodology is implemented to diagnose the volume of a tumor and to check the cancer infection as to how far and to what extend the problem has developed. Staging helps the doctor to plan the best treatment for an individual patient. One the best and most accurate methods to define the cancer is TNM system. The TNM system is implemented to identify the position or condition among four stages.

Stage 1 is considered to be small and less effected inside the breast (Fig.11.3).

Stage 2 is considered to be category where the surrounding tissues are not affected in the breast, but larger than Stage 1 and moved to lymph nodes closer to tumor, even growth has not started in remaining organ in the human body (Fig.11.4).

Stage 3 compared to Stage 1 and 2 the cancer is bit larger. In this stage the cancer cells have started moving into surrounding tissues and even to the lymph nodes in the area (Fig.11.5).

Stage 4 is at high risk where the cancer cell has moved to rest of the organs of the body. Also termed has secondary or metastatic cancer (Fig.11.6).

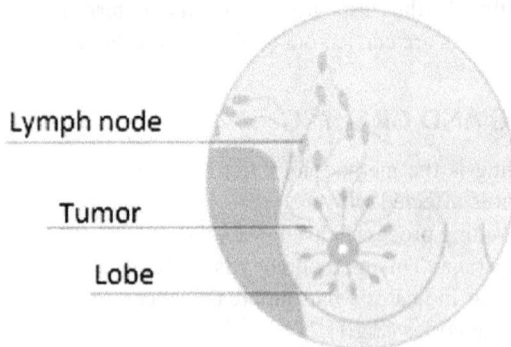

FIGURE 11.3   Stage 1 of tumor size.

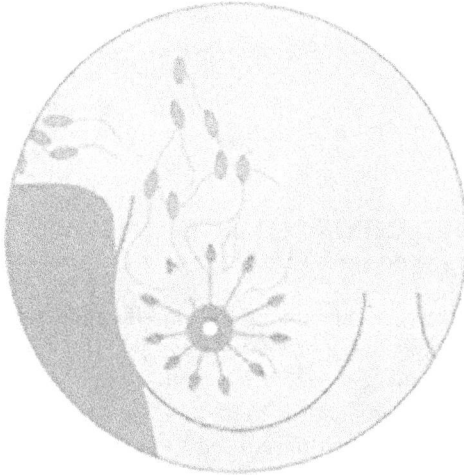

FIGURE 11.4   Stage 2 of tumor size.

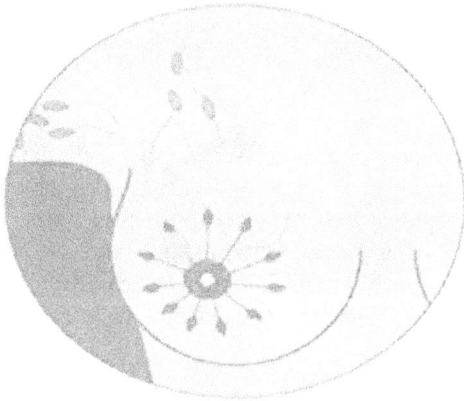

FIGURE 11.5   Stage 3 of tumor size.

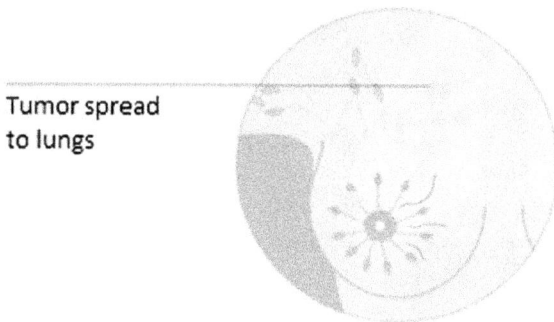

Tumor spread
to lungs

FIGURE 11.6   Stage 4 of tumor size.

These are special kind of tests that are invented those that are used by the health care professionals in diagnosing the breast cancer. It is not necessary that all cases need these tests. Accuracy in diagnosis depends on the situations. All these are considered for non-cancerous the benign stage which changes as the pathologist observes under the microscope [1].

11.7 EVALUATING AND IMPLEMENTING NAIVE BAYES ALGORITHM

In the world of data science one among the classification algorithms is the naive Bayes algorithm, which is based on Bayes' theorem. The naive Bayes is a popular method for creating statistical predictive models. This algorithm is most commonly used for problems related to prediction for ease of implementation and usage. This classification algorithm is used to analyze the relationship between the attribute and various classes to determine the prediction or probability of the problem.

According to this study for the given patient record, every record is grouped based on the conditions as fit ("Recurrence event") or unfit ("N0-Recurrence event") for class.

The dataset is classified into two categories, the feature matrix and the response vector.

- The feature matrix represents each vector (row) of the datafile and each vector contains values which are dependent. Based on the dataset considered for the study, features are age, menopause, tumor-size, inv-nodes, node-caps, deg-malig, breast, breast-quad, and irradiat.
- The response vector represents the actual value of the variable for each row in the feature matrix. Similarly, the name of the variable in the class is the class itself.

11.7.1 ASSUMPTIONS

The naive Bayes algorithm assumes that each feature in the dataset is independent and contributes equally to the result.

Based on the relationship between features in the dataset, the assumptions can be understood as:

- Each feature is independent of each other.
- Equal importance is given to each feature.

11.7.2 BAYES' THEOREM

In terms of theoretical probability and event, given the probability of another event the Bayes theorem determines the probability for an event occurring. A mathematical representation of Bayes' theorem:

$$P(A/B) = (P(B/A)P(A))/P(B)$$

Where A and B are events and P (B) $\neq$ 0

P (A/B): Possibility of action A occurrence assuming that event B is true.
P (B/A): Possibility of action B occurrence assuming that event A is true.
P (A) and P (B) are actual possibilities that occur, which are independent of each other.

11.8 AREAS UTILIZATION NAIVE BAYES ALGORITHMS

- **Identifying and Solving Worldwide Problems:** In the fast-moving world, naive Bayes is an learning classifier that can be used for real-world prediction.
- **Various Leveled Classes Analysis:** This analysis method is absolutely suited for various leveled classes analysis or prediction. Predictions are made on the different feasibility of various classes of destination variable.
- **Text Regulation/Spam Refinement/Emotional Analysis:** Naive Bayes classifiers are commonly implemented in text regulation for better outcomes in various level class problems. They give a better output rate when compared to other popular algorithms. This algorithm is popularly used in Spam refining to extract spam e-mail and also for emotional analysis in the field of social media analysis, to analyze the pros and cons of customer sentiment.
- **Suggestion System:** Recommendation systems are developed using two different approaches like the naive Bayes algorithm and the collaborative filtering method, in which machine learning methods and data mining techniques are applied to extract unknown information and determine whether an end user would accept a given resource or not.

11.8.1 BUILDING PRIMARY MODULES USING NAIVE BAYES

The Python library scikit is used in the research process to build a naive Bayes model. The scikit-learn (sklearn) library includes three different forms of the naive Bayes model:

- **Gaussian:** This model is implemented in classification techniques. It makes the assumption that characteristics follow a normal distribution.
- **Multinomial:** This model is implemented for discrete counts. For instance, consider, a classification problem. In this case Bernoulli trials are considered. In the example, in addition to "word reappearing in the document," it is necessary to "keep track of how often the word reappears in the document." Another example is "number of times resultant number x_i is determined in n trials".

- **Bernoulli:** The binomial model is effective when feature vectors are binary (i.e., zeros and ones). A simple, real-time example would be text analysis, i.e., a pool of content condition. In this case 1s and 0s are "word reappears in the document" and "word does not reappear in the document."

11.9   SYSTEM WORKFLOW ARCHITECTURE

11.9.1   EXPLORATORY DATA ANALYSIS

Jupiter Notebook is used to work on the dataset by importing necessary libraries and by importing the dataset to Jupiter Notebook

11.9.2   SPLITTING THE DATASET

The data is subdivided into training data and test data. The training set usually poses predicted outcome on the basis of which the program learns to make predictions on a new data file. The test data (or subset) is used to test the predictions made by the program and this is done using sklearn library in Python using the train_test_split method.

11.9.3   DATA WRANGLING

This step loads in the data, checks for cleanliness, and then trims and cleans the given dataset for analysis.

11.9.4   DATA GATHERING

The data gathered for predicting patient situations is subdivided into a training set and a test set. Broadly, 7:3 classification is implemented to categorize the training dataset and the test dataset. The module designed by programmers using the naive Bayes algorithm is evaluated using the training set and, based on the outcome efficiency, the test dataset is applied for further predictions.

11.9.5   PREPROCESSING

The obtained records probably have some missing records resulting in data inconsistency. To achieve an accurate result, the data needs to be preprocessed. The errors have to be neglected and data modifications need to be carried out. Based on the interrelationship among aspects, it was observed that the significantly individual aspects include TNM, stages, grade, and age, which is the strongest among all.

11.9.6 BUILDING THE CLASSIFICATION MODEL

For predicting breast cancer, a decision tree algorithm is most appropriate because an appropriate outcome can be expected in classification as follows:

- It is efficient when reprocessing errors in the data, irrelevant variables, and a combination of all type of continuous, categorical, and discrete variables.
- It results in a huge amount of expected mistakes; those that are proved to be improper through many tests.

11.9.7 TRAINING THE DATA

- In step 1, import the iris data set that is available ready to incorporate in the sklearn module. Raw data that is available usually constitutes different varieties of variables.
- For example, import any method and train_test_split class from sklearn and NumPy module to be incorporated in this kind of program.
- To inculcate load_data() method in data_dataset variables further, a train_test_split method is incorporated to classify the data into Training and test data. The N prefix is the variable that explains the feature values and Y prefix represents target values.
- As explained earlier, the raw data is classified into training and test data, probably in the ratio of 67:33/70:30. Later this data is incorporated in the developed module.
- Finally, the training data set is classified used to train the module for further analysis and to produce the best result.

11.9.8 TESTING THE DATASET

- In this phase, various aspects of features in NumPy array are represented using variable 'n' to report the different aspects of the feature. To perform this the prediction method applies NumPy array works on entering data and splits out the desired value as a result.
- Finally, the identified target value is obtained to be 0 as a result. Furthermore, to analyze the result, score the results as a ratio of the number of identifications obtained as absolute to the overall identification obtained. Finally, determine an evaluation score which evaluates the real values and the predicted values.

11.9.9 SYSTEM WORKFLOW

According to this diagram when the user enters the data, errors in the data are first removed and then the algorithm is applied to determine the TNM levels and grade cells that further decide the patient stages (Figure 11.7).

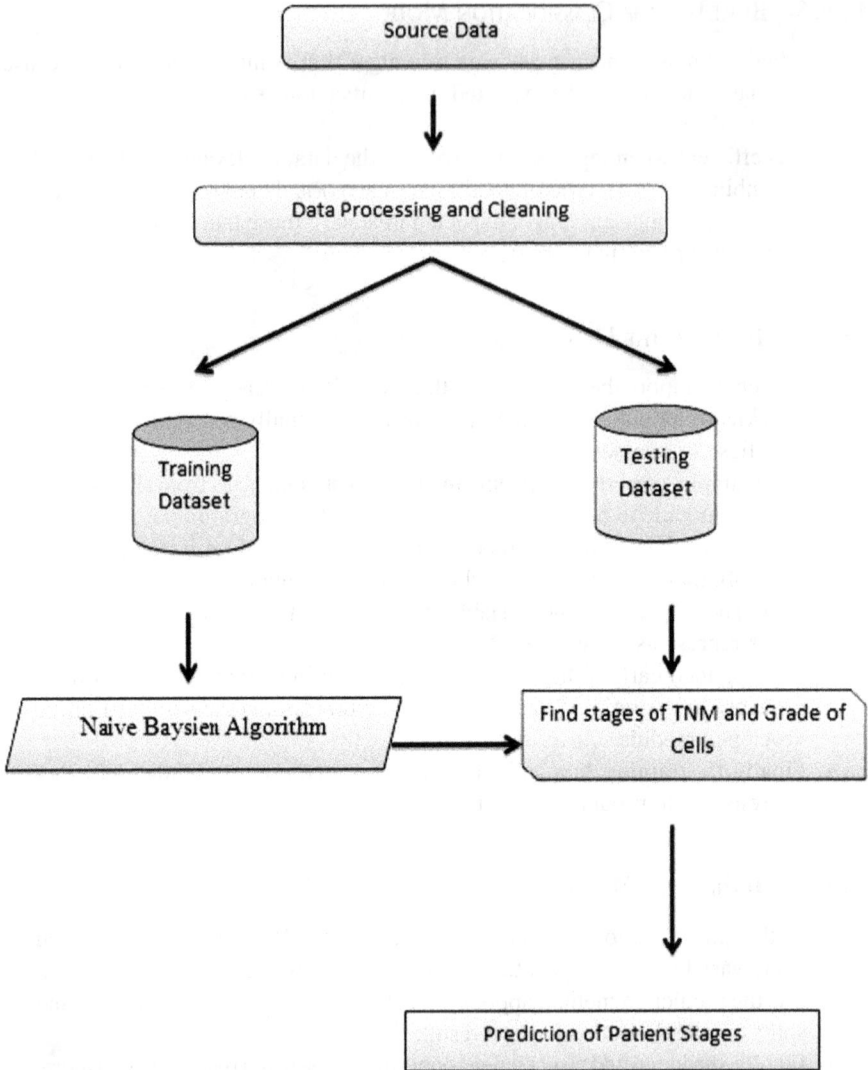

FIGURE 11.7   System workflow diagram.

11.10   IMPLEMENTATION OF NAIVE BAYES ALGORITHM USING ANACONDA SOFTWARE

The algorithm is implemented in the dataset to determine the stage of the breast cancer in the affected patient by considering the attribute as important aspect for analysis.

Now, according to the given dataset, Bayes' theorem can be applied in following way:

$$\mathbf{P(Y/N) = (P(N/Y)P(Y))/P(N)}$$

Where Y is variable class and N is feature vector which is dependent of size *n* such that:
N = (n1, n2, n3............)

Basically, according to the study, P (N|Y) represents, the probability of "Not class" given the patient conditions attributes are TNM, stage, and age.

Generally, one of the types of methodology that comes under the naive Bayes is the Gaussian naive Bayes. It is particularly used in cases where the character of the variable is regular. It works on the assumption that all the characters are of Gaussian distribution, which is also called the normal distribution.

For example, applying the Gaussian naive Bayes classifier using sklearn:

```
# Bundle the iris dataset
Import pandas as p
iris = p.reannid_csv("dataset.csv")

# storing the feature matrix (N) and response vector (Y)
N = iris.data
Y = iris.target

# Dividing N and Y as training and testing datasets
from sklearn.model_selection import train_test_split
N_train, N_test, Y_train, Y_test = train_test_split(N, Y,
test_size=0.4, random_state=1)

# program is trained using training dataset
from sklearn.naive_bayes import GaussianNB
gnb = GaussianNB()
gnb.fit(N_train, Y_train)

#prediction made on testing data
Y_pred = gnb.predict(N_test)

# correlating definite response values (Y_test) with
identified response values (Y_pred)
Y from sklearn import metrics
print("Gaussian Naive Bayes model accuracy(in %):", metrics.
accuracy_score(y_test, y_pred)*100)
```

11.11 EVALUATING THE ACCURACY OF SOFTWARE MODULES

11.11.1 CONFUSION MATRIX

A confusion matrix is a tabular representation that is commonly used to **evaluate the performance of a classification model** (or "classifier") on a given test data for which the true values are identified earlier. The confusion matrix is not too complex to understand, but the related terms used are bit confusing.

11.11.2   Basic Terms

True Positives (TP): A person who will not pay is predicted as a defaulter. These are values that are predicted correctly called positive values; as such the outcome of both the real and identified class is positive.

   True Negatives (TN): A person who defaults is predicted as a payer. These are values that are predicted correctly as negative values; as such the outcome of both the real and identified class is negative.

   False Positives (FP): A person who will pay is predicted as a defaulter. In this case the real class value is negative and the identified class value is positive.

   False Negatives (FN): A person who will default is predicted as a payer. In this case the actual class value is yes but the predicted class value is no.

11.11.2.1   Comparing the Algorithm with Prediction in the Form of Best Accuracy Result

Many machine learning algorithms are developed by many programmers; it is necessary to compare these machine learning models. Automated testing methods are implemented to correlate different machine learning algorithms in Python with scikit- learn. The automated analysis testing methods are used as examples of many different kinds of algorithms. Performance varies in each module. Implementing resampling methods is like cross validation; it can determine how accurately each model functions on unknown data. Looking at data from a different angle while selecting it, the same methodology is applicable to model selection. Different accuracy calculation methods are incorporated to determine the best selection.

   In this section you will see the procedure to carry out the same in Python sklearn. The basis of a valid correlation of machine learning algorithms is to ensure that each algorithm is examined in the same procedure on the equal data and this can be accomplished by inculcating every algorithm to be calculated based on an efficient automated test methods.

11.11.2.2   Prediction Result by Accuracy

Logistic regression determines a value using linear equation with independent predictors. The predicted amount can be any combination of false infinity to true infinity [5]. The required outcome of the algorithm is to be classified variable data. Higher accuracy when predicting the result is determined with a logistic regression model by comparing the best accuracy.

$$\text{True Positive Rate (TPR)} = TP / (TP + FN)$$

$$\text{False Positive Rate (FPR)} = FP / (FP + TN)$$

Accuracy: Accuracy is defined as the proportion of the total number of predictions that is correct; alternatively, how often the model correctly predicts defaulters and non-defaulters overall.

11.11.2.3 Accuracy Calculation

$$\text{Accuracy} = (\text{TP} + \text{TN}) / (\text{TP} + \text{TN} + \text{FP} + \text{FN})$$

Accuracy is the most effective performance method used to determine the ratio of correct predicted observation to the total observations. It is a common assumption that if there is high accuracy then the model is considered to be best. This assumption is correct; accuracy is a high performance measure when there is a symmetric datasets in which values of false positive and false negatives are almost similar.

Precision: Precision is the proportion of positive predictions that are actually correct. (When the model predicts default: how often is correct?)

$$\text{Precision} = \text{TP} / (\text{TP} + \text{FP})$$

Precision is calculated by dividing correctly predicted positive observations with total predicted positive observations.

Recall: Recall (sensitivity) is the proportion of positive observed values correctly predicted.

$$\text{Recall} = \text{TP} / (\text{TP} + \text{FN})$$

Recall is calculated by dividing correctly predicted positive observations with all observations in actual class—yes.

F1 Score is calculated by determining the weighted average of precision and recall. Therefore, this calculation takes both false positives and false negatives into consideration. Learning this is not as simple a task as accuracy, but F1 is actually more effective than accuracy, mainly in the case of uniform class classification. Accuracy is effective only when false positives and false negatives pose same cost. If false positives and false negatives vary, the best choice is to consider both precision and recall.

General Formula:

$$\text{F} - \text{Measure} = 2\text{TP} / (2\text{TP} + \text{FP} + \text{FN})$$

F1-Score Formula:

$$\text{F1 Score} = 2*(\text{Recall} * \text{Precision}) / (\text{Recall} + \text{Precision})$$

11.11.3 RESULTS OF TESTING USING CONFUSION MATRIX

Results are analyzed using confusion matrix to measure the performance of machine learning algorithm. Here, the row represents the actual class and the column represents the predicted class. The fields in the confusion matrix represents True positive, True negative, False positive, and False Negative (Table 11.1).

TABLE 11.1
Confusion Matrix

	Predicted		
	Negative	Positive	Total
Actual			
Negative	59	0	59
Positive	0	25	25
Total	59	25	84

11.12 CONCLUSION

Breast cancer is a disease caused by abnormal cells growing in the breast numerously. There are various types of breast cancer. The type of breast cancer is determined by which cells in the breast change into cancer. Cancer can affect different areas of the breast. Most breast cancer starts in the ducts or lobules. Breast cancers can move outside the breast by blood vessels and lymph vessels. When breast cancer moves to other organs of the body, it is considered to be metastasized.

These cells usually develop into tumor cells that are visible on an X-ray or felt as a lump. The tumor is malignant (cancer) if the cells can develop into (invade) surrounding tissues or move (metastasize) to different organs of the body. Breast cancer is most common among women, but rarely men can get breast cancer.

Machine learning is technology in the field of AI that encapsulate a wide variety of statistical, probabilistic, and optimization techniques that enable computer models to develop expertise from previous examples. This is especially capable and well-suited to use in medical applications whose main factors depend on complex proteomic and genomic measurements. In short, machine learning is commonly used in cancer diagnosis and detection. Machine learning is now also used in cancer prognosis and prediction.

In this chapter, the machine learning classifier naive Bayes classifier is implemented to determine the stage of a breast cancer patient and to grade the cell size. The TNM system is used to explore the stage. Accurate performance of the naive Bayes algorithm is evaluated by calculating the accuracy, specificity, sensitivity, and F1 score as shown in Table 11.2 [5].

TABLE 11.2
Performance of Test Data

Method	Accuracy	Specificity	Sensitivity	F1 Score	Recall	Precision
Naive Baysien	100	100	100	100	100	100

REFERENCES

1. Ganesh N. Sharma, Rahul Dave, Jyotsana Sanadya, Piush Sharma, and K. K. Sharma, "Various Types and Management of Breast Cancer: An Overview," Journal of Advanced Pharmaceutical Technology & Research, vol. 1(2), Apr-Jun 2010, PMCID: PMC3255438, PMID: 22247839.
2. Osvaldo Simeone, "A Very Brief Introduction to Machine Learning With Applications to Communication Systems," Institute of Electrical and Electronics Engineers, vol. 4(4), Nov 2018, doi: 10.1109/TCCN.2018.2881442
3. Yi-Sheng Sun, Zhao Zhao, Zhang-Nv Yang, Fang Xu, Hang-Jing Lu, Zhi-Yong Zhu, Wen Shi, Jianmin Jiang, Ping-Ping Yao, and Han-Ping Zhu, "Risk Factors and Preventions of Breast Cancer," International Journal of Biological Science, vol. 13(11), Nov 2017, doi: 10.7150/ijbs.21635, PMCID: PMC5715522, PMID: 29209143.
4. Jieun Koh, and Min Jung Kim, "Introduction of a New Staging System of Breast Cancer for Radiologists: An Emphasis on the Prognostic Stage", Korean Journal of Radiology, vol. 20(1), 27, Dec 2019, doi: 10.3348/kjr.2018.0231, PMCID: PMC6315072, PMID: 30627023.
5. D. R. Umesh, and C. R. Thilak, "Predicting Breast Cancer Survivability Using Naïve Baysien and C5.0 Algorithm," International Journal of Computer Science and Information Technology Research, vol. 3(2), 802–807, April-June 2015, ISSN 2348-1196 (print), ISSN 2348-120X (online). Available at: www.researchpublish.com.

12 Deep Networks and Deep Learning Algorithms

Tannu Kumari and Anjana Mishra
C.V. Raman Global University

CONTENTS

12.1 INTRODUCTION

The most advanced organ in a person's body is his brain. It decides the manner in which we see, hear, smell, taste, and feel. The brain allows us to store dreams, various experiences, emotions, and also our dreams. Without it we would be mere common creatures, deprived of doing any different tasks than the simplest. In our whole body the part which is responsible for making us think before doing any task and which provides us intelligence is the brain. It weighs only three pounds, approximately one and a half kilograms, which is sufficient to solve the most complex

FIGURE 12.1   Dogs or chicken.

problems that our super equipped computers can't solve or find difficult to solve. After the birth of a human child in only in few months the baby starts recognizing the faces of their parents, starts identifying objects, and even start producing different voices. In their early stages they start locating objects even after getting blocked and start relating different kinds of things by their sounds. After growing to their early school days they start learning about grammar and start getting familiar with many different kinds of words.

Here is proof that shows that we humans sometimes lose on classification tasks despite having been trained for millions of years! (Figures 12.1–12.5).

12.2   DEEP LEARNING

Since early days humans have always kept pushing forward and inventing new machines that would have brains similar to humans, such as robots serving as helpers in our household activities and devices like microscopes which are able to identify disease on their own. To do this we have to design artificial intelligent programming machine, which requires us to work on some of the most complex computer-related types of problems with which we have never struggled with before. These are problems that our mind can solve easily in fractions of seconds. In order to handle these

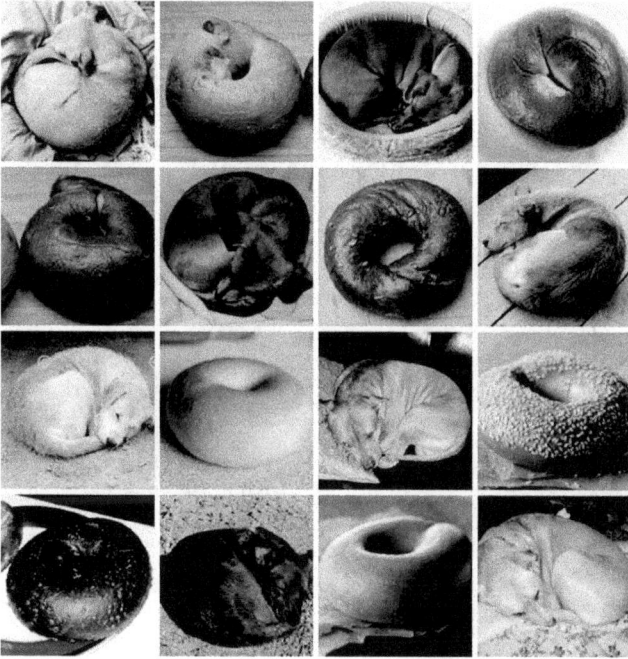

FIGURE 12.2 Dogs or donuts.

FIGURE 12.3 Dogs or mop.

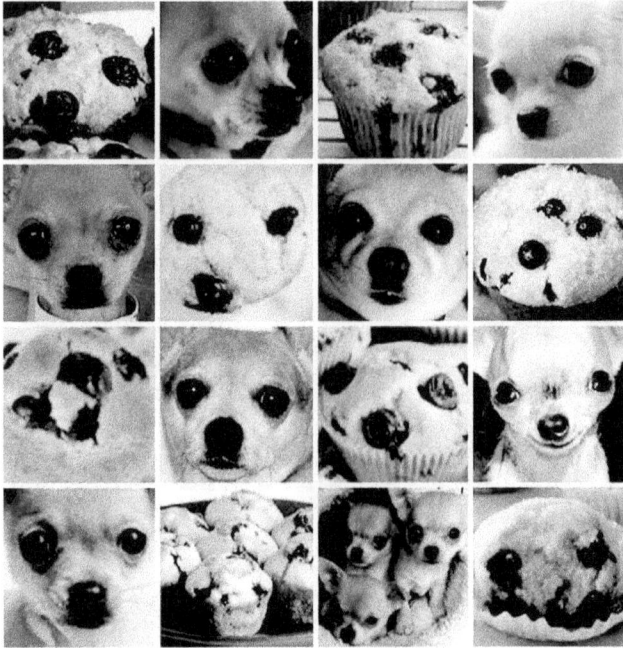

FIGURE 12.4 Dogs or cookies.

FIGURE 12.5 Parrot or something else?

tasks, we need to develop a very different way to program a computer, with the help of techniques which have been developed in the very past decade. Techniques called deep learning generally refer to the network which is active, an area of artificial intelligence (AI) in machines.

12.2.1 Why Deep Learning

There is no doubt that machines are really fast; they can do calculations in seconds that a human brain would take several minutes to do. Also, their performance and accuracy are unmatchable. But what if a machine or a computer is given a handwritten text to read and calculate? Take, for example, Figure 12.6.

The computer would not be able to recognize the alphabets or digits as efficiently as we can. So, this is an example of a need for deep learning. Actually, deep learning can be said to be a subset of the generalized field of AI that is called machine learning (ML), based on the principle of learning from example. ML works by a very different method: it doesn't follow a set of long rules for solving problems, it is given a task according to which it solves different problems following instructions of algorithms for modifying models whenever it makes any kind of mistake. We are currently working on very well-suited models that could solve any problem very accurately within a very short period of time (Buduma, 2017).

12.2.2 The Perceptron (Neuron)

The main part of our brain is referred to as a neuron. As we know, the size of a grain of wheat is very small, similarly a very small part of brain has more than 10,000 particles of neurons in which approximately 5,000 links are established in between the neurons. It is this big biological connection that allows us to explore the huge

FIGURE 12.6 Handwritten notes.

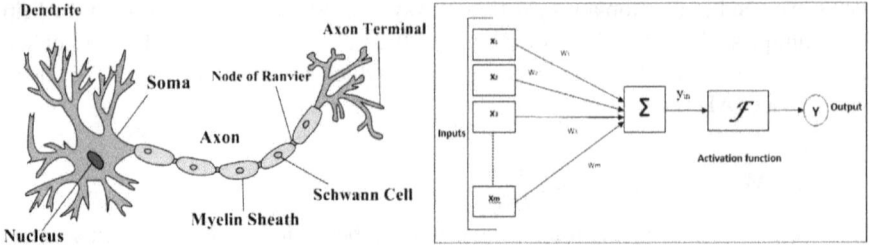

FIGURE 12.7   The neuron.

world around us in a very broad manner. The precept for us is to use a similar structure to build out ML blocks that could help us solve any type of problem. Internally the neuron is made in such a way that it could get data from other neurons, process the data in a very unique format, and transfer the output to other micro parts such as cells. Figure 12.7 illustrates the process.

Inputs are received by the neurons through tentacle-like bodies known as dendrites. All of these links are randomly either strong or weak on the basis of how frequently they are used, and the ability of every link shows how much the inputs have contributed to the output of the neurons. As soon as the primary attributes of the simultaneous connections are recognized, the inputs are made to sum up together. This summed up thing is then converted to a different new signal, which is then transmitted along the axon of the cell and then sent to all the other neurons (Buduma, 2017).

12.3   NEURAL NETWORKS

The artificial network is a prototype modeled by the architecture of the human brain. In conventional prototypes of the human brain, a huge number of simple computers work together, through which our human brain can proceed with high-level computational algorithms. Super computational architectures are designed along this computational paradigm. Attempts to learn along with these networks were introduced during the twentieth century. Defining a neural network includes nodes and links, where it is graphed with nodes (neurons) and edges (links between the nodes). Every node gets input as a summation of the weighted output nodes connected to the incoming links (Apostolou and King, 1999; Ben-David, 2014).

12.3.1   Network Topologies

Given the composite elements of the neural design, now follows a clear idea of the general designs (topologies) of neural type of networks, i.e., to build networks having these elements. All the topologies shown here in this text are demonstrated by a map and its Hinton diagram; light gray fields are used to show the dotted weights, dark ones for solid weight. In the Hinton diagram the dotted weights, which have been added for clarity reasons, could not be found there. To clarify the connections between the neuron lines and neuron columns, there is a small arrow in the upper-left cell inserted by me.

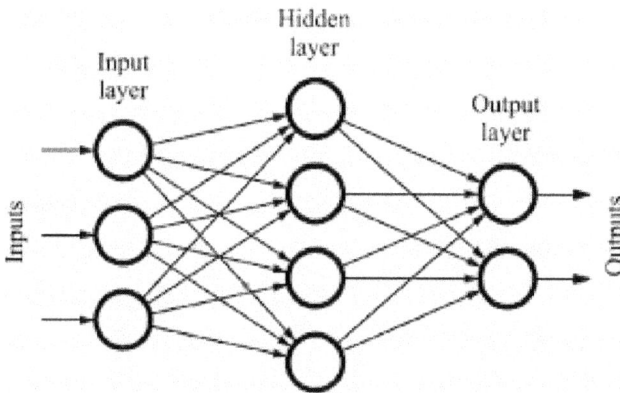

FIGURE 12.8   Feed forward neural network.

12.3.2   Feedforward Neural Networks

While single neurons are much stronger than linear perceptron, they are not so expressive that they can solve the difficult problems of learning. There is a reason why our brain is made up of so many neurons. Take, for example; the ability to differentiate between handwritten digits, which is nearly impossible for a single neuron. In order to tackle these types of complex activities we have to take our ML model a step ahead. Our brain is made up of multiple layers of neurons. In our brain the model responsible for most of the intelligence activities is the cerebral cortex, which is made up of six layers. The information flows from each layer unless the sensory input is converted to logical understanding. The lowermost part of the brain, known as the visual cortex, gets raw visual data from the eyes. Each layer processes the information and keeps passing it to the next layer until it reaches the last one. Figure 12.8 shows more details of these layers (Buduma, 2017).

Layers and connections for every layer are present in a feedforward network. Here the first thing that we are going to look at (although different topologies are going to be used later) is feedforward network which are networks. Grouping neurons is done in layers which are: an input layer, n hidden processing network of layers (not visible through the out-layers side, which is why the neurons are also said to be hiding), and an output layer. Each neuron in one layer in a feedforward network has just the direct connections with the next layer of neurons (towards the output layer). Figure 12.9 represents the permitted connections (solid lines) for a feedforward network. Preventing a collisions of names is handled by referring to the output neuron as Ω (Kriesel, 2005).

12.3.3   The Artificial Neuron

The biological neuron is simulated in an artificial neural network (ANN) by an activation function. In classification tasks (e.g., identifying spam e-mails) this activation function must have a "switch on" characteristic – in other words, once the input is

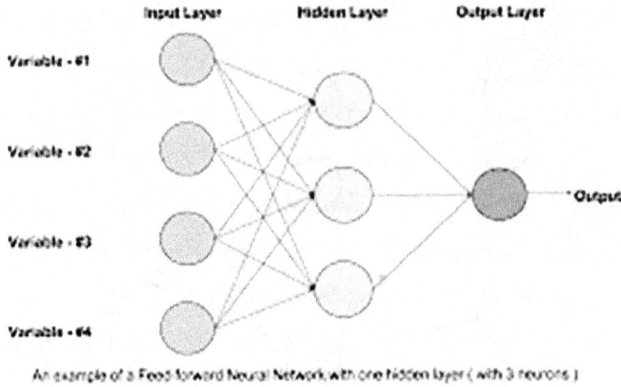

An example of a Feed forward Neural Network with one hidden layer (with 3 neurons)

FIGURE 12.9 Permitted connections for feed forward networks.

greater than a certain value, the output should change state i.e., from 0 to 1, from −1 to 1, or from 0 to >0. This simulates the "turning on" of a biological neuron. A common activation function that is used is the sigmoid function, which looks like this:

$$f(z) = \frac{1}{1 + \exp(-z)}$$

As can be seen in Figure 10.9, the function is said to be "activated," i.e., it moves from 0 to 1, when the input x is greater than some specific value. The sigmoid function isn't a step function, however, the edge is "soft," and the output doesn't change instantly. This means that there is a derivative of the function and this is important for the training algorithm (Thomas, 2017).

12.3.4 GRADIENT DESCENT

Let's look at how we can reduce the doubled mistakes on every training instance by looking into the problem simply. Assume that our straight neuron has only two weights, $w1$ and $w2$. After this, we could consider an instance of a 3D space where the horizontal dimensions correspond to the weights $w1$ and $w2$ and the linear dimension refers to the value of the error equation Z. Considering the current area, points in the linear plane refer respectively to diverse changes of the respective weights, and the length at those points refer to the error that has occurred. If the errors are considered we create over all considerable weights, we get an area in this 3D area, specifically, a parabolic surface as shown in Figure 12.10.

One can also simply see this area as a cluster of elliptical curves, where the least error is at the center. In this setup, we are working in a two-dimensional plane where the dimensions correspond to the two weights. Contours correspond to settings of $w1$ and $w2$ that evaluate to the same value of E. The closer the contours are to each other, the steeper the slope. In fact, it turns out that the direction of the steepest descent is always perpendicular to the contours. This direction is expressed as a vector known

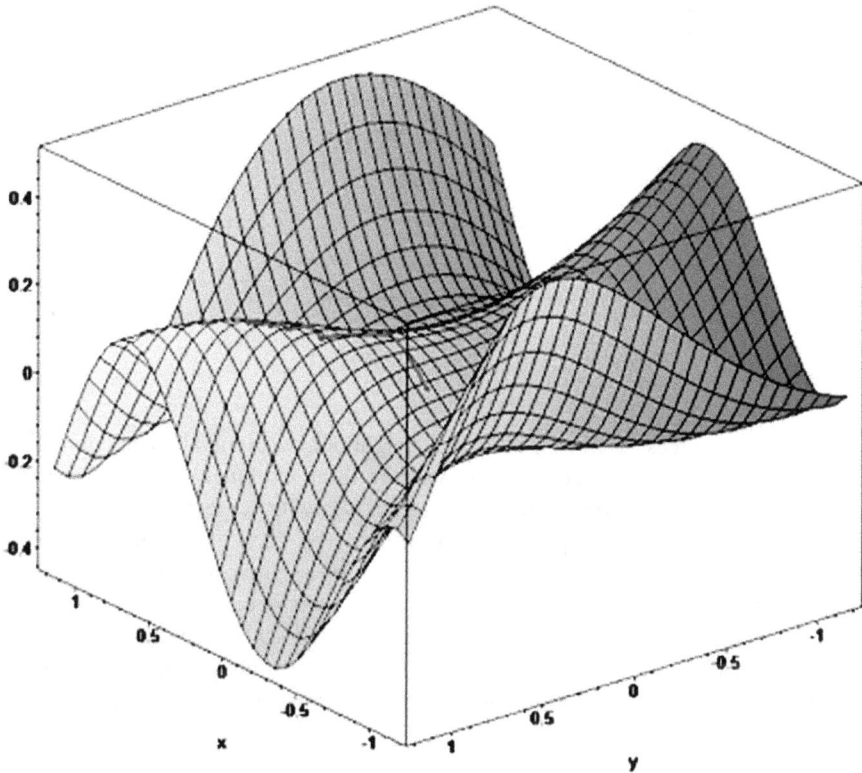

FIGURE 12.10   Gradient descendent.

as the gradient. Now we can develop a high-level strategy for how to find the values of the weights that minimizes the error function. Suppose we randomly initialize the weights of our network so we see all of us on the linear area. By evaluating the gradient at our current position, we can find the direction of steepest descent, and we can take a step in that direction. Then we'll find ourselves at a new position that's closer to the minimum than we were before. We can re-evaluate the vector of steepest descent by taking the gradient at the new position and taking a step in this new direction. It's easy to see that, as shown in Figure 12.11, on considering this plan we would slowly get to the point of least error. This algorithm is known as gradient descent (Buduma, 2017).

12.3.5   THE BACKPROPAGATION ALGORITHM

The backpropagation algorithm was pioneered by David E. Rumelhart, Geoffrey E. Hinton, and Ronald J. Williams in 1986. Can we guess the centralized idea behind the evolution of the backpropagation algorithm? We do not realize what the unseen entities might be working on, all we can work on is to calculate how rapidly the error changes as we change an unseen activity. From there, we can figure out how

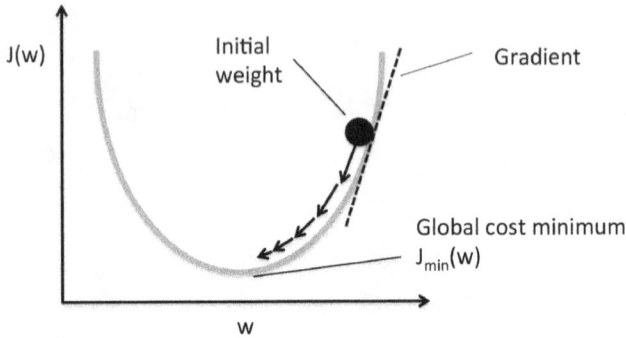

FIGURE 12.11 Backpropagation.

fast the error changes when we change the weight of an individual connection. This is the only way that we're going to be working in an extremely high-dimensional space. We begin by computing the error derivatives with respect to a single training example. Numerous output units are affected by each of the hidden things. Numerous other changes on the error in a very informative way will have to be populated by us. Programming dynamically will be one of our methods. As soon as the derivatives of error are known to us for a single hidden unit layer, we will use it to calculate the byproduct of error activities for invisible units; it is little bit easy for getting out the derivatives of the erroneous part of the weights leading to hidden unit (Buduma, 2017) (Figure 12.12).

12.3.6 TENSORFLOW AND ITS USE

TensorFlow is a Python library that allows users to convey arbitrary computations as a graph of data flow rate. Nodes in this graph represent mathematical operations,

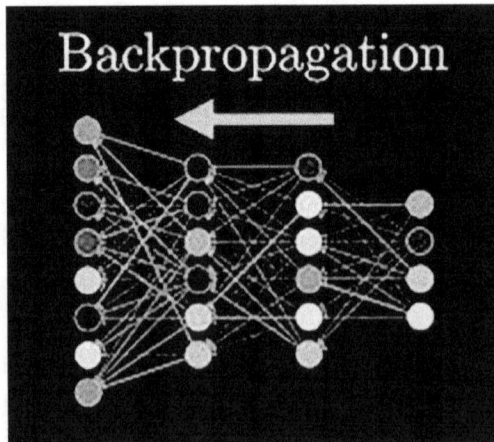

FIGURE 12.12 TensorFlow operation.

while edges represent data that is communicated from one node to another. Data in TensorFlow is represented as tensors, which are multidimensional arrays (representing vectors with a 1D tensor, matrices with a 2D tensor, etc.) (Buduma, 2017).

Apart from TensorFlow, there are other libraries that have come up over the decade for building deep neural networks. These include Theano, Torch, Caffe, Neon, and Keras.

12.4 NEURONS IN HUMAN VISION

The visual ability of human beings is amazingly advanced. We can detect anything within our range of sight without any kind of disturbance in a very short period of time. It's not just can do nomenclature things we are seeing through, we are also able to look at the depth, differentiate contours, and discriminate the things from their background. We know that our eye receives pixels of data in the raw format of color data, it is our brain which processes that raw data and gives out more meaningful results such as curves, lines, and directions that tell us, for example, that we are seeing a pet dog. Neurons are primary to mankind's senses. There are some special types of neurons that help humans receive light data from their eyes. The data collected from the light is processed beforehand and then transferred to the visual cortex of the brain, and afterward analyzed for its end task. It is the neurons that are responsible for all of the activities taking place. One conclusion from this is that initially this will make lots of sensory activities for computers possible by expanding our neural-connections prototype for building out the best vision applications. To build out a better prototype of deep learning for problems related to images we would be using our sense of perception of our sense of sight. Before diving in, let's look at traditional approaches for analyzing an image and why they often failed for extra-meaningful types of data such as shapes, lines, and curves which could, for instance, be indicating that we are seeing a pet dog (Bishop, 2006; Buduma, 2017).

12.4.1 THE PREDICTORS: DECISION TREES

The predictor, h: X → Y, predicts the tag considered along the entity X by traversing through the leaf to root on a tree. In simple terms, the focus is provided to the binary differentiation, that is, Y = {0, 1}; however, these trees can be used for various predictions also. Along every node on the root to leaf area, the successor node is selected on the capability of splitting input area (Ben-David, 2014).

12.4.2 LEARNING LOWER-DIMENSIONAL REPRESENTATIONS

We observed that the convolutional architecture uses a simple argument. The greater our input vector, the greater our model. Large models with a number of parameters are expressive, but they're also increasingly data hungry. This means that without sufficiently huge volumes of training data, we will likely over fit. Convolutional architectures help us cope with the curse of dimensionality by lowering the number of parameters in our models without necessarily removing expressiveness. Regardless, convolutional networks still require large amounts of labeled training data. And for

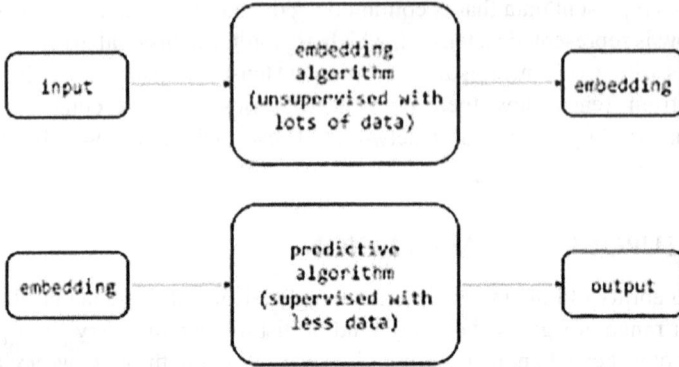

FIGURE 12.13 Convolutional neural networks.

many problems, labeled data is scarce and expensive to generate. Our goal in this section would be to develop efficient learning models in situations where labeled data is scarce but wild, unlabeled data is plentiful. We'll approach this problem by learning embeddings, or rather low-dimensional representations, in an unguided fashion. Because these unguided models allow us to offload all of the heavy lifting of automated feature selection, we can use the generated embeddings to solve learning problems using comparatively smaller models that require less data. This process is summarized in Figure 12.13 (Buduma, 2017).

12.4.3 CONVOLUTIONAL LAYERS

A convolutional neural network is a class of deep neural networks, most commonly applied to analyzing visual imagery. They are also known as shift invariant or space invariant artificial neural networks. Figure 12.14 depicts Convolutional neural network.

12.4.4 REINFORCEMENT LEARNING

Reinforcement learning refers to the psychology and the cognitive science that describes the mechanism of learning on its own by analyzing or practicing a certain task and learning from its own mistakes. The problem is that there is no method of

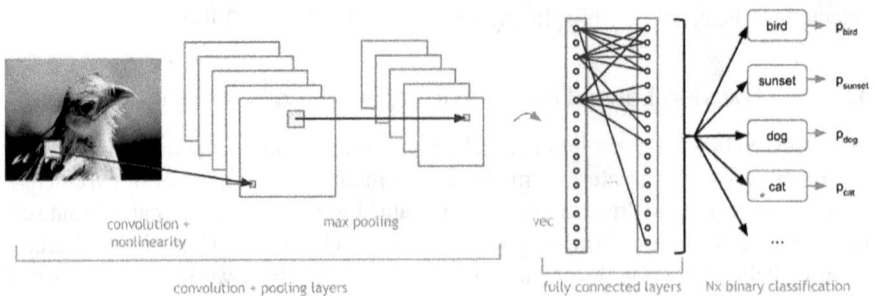

FIGURE 12.14 Representing convolutional networks.

learning that truly describes the functions that we have to perform: the thing that we get at last is the overall result, like whether we won that game or not. And what are the chances of winning that game? But there is no exact answer for the intermediate steps. The aim of reinforcement learning is basically to maintain the feeling of having suitable examples of maximizing the benefits one can get from a specified task. Let's take the example of a tennis player who is trying to keep his success rate in athletics at its peak for the long run by using complex moves and mastering the curvy trajectories in 3D space, taking into consideration the direction of the wind, the series importance, and all other factors. Coming directly to the point, the main thing is that if we receive only a little feedback, progress is very slow as because reinforcement learning means continuous trial mechanism until the incoming error is stopped (Kriesel, 2005).

12.4.5 RECURRENT NEURAL NETWORKS

Recurrent neural networks (RNNs) were first introduced in the 1980s, but have regained popularity recently due to several intellectual and hardware breakthroughs that have made them tractable to train. RNNs are different from feedforward networks because they leverage a special type of neural layer, known as recurrent layers, that enable the network to maintain its state between uses of the network. Figure 12.15 illustrates the neural architecture of a recurrent layer. All of the neurons have both (1) incoming connections emanating from all of the neurons of the previous layer, and (2) outgoing connections leading to all of the neurons in the subsequent layer. We observe here, however, that these aren't the only connections that neurons of a recurrent layer have. Unlike a feedforward layer, recurrent layers also have recurrent connections, which propagate information between neurons of the same layer. A fully connected recurrent layer has information flow from every neuron to every other neuron in its layer (including itself). Thus, a recurrent layer with r neurons has a total of $r2$ recurrent connections (Buduma, 2017; Schmid Huber, 2015).

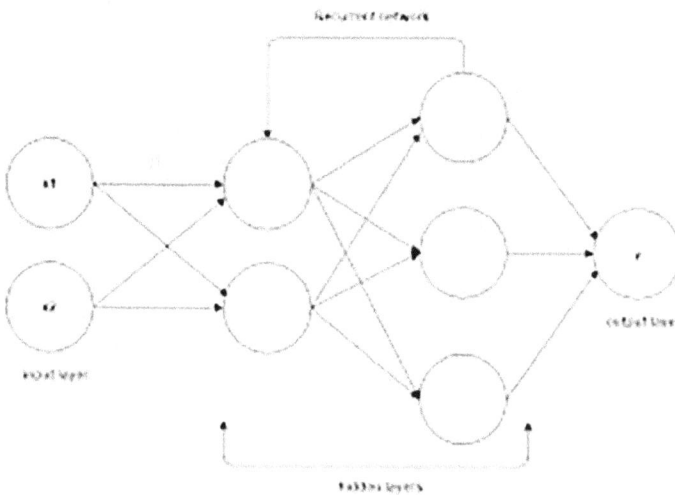

FIGURE 12.15   Recurrent neural networks.

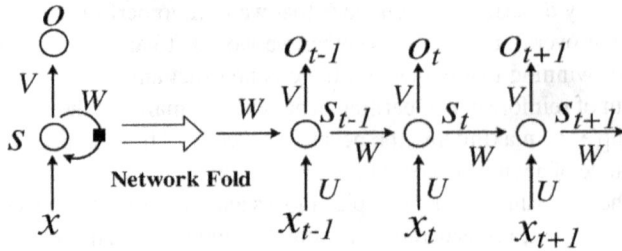

FIGURE 12.16 Unrolling the RNN through time.

To better understand how RNNs work, let's explore how one functions after it's been appropriately trained. Every time we want to process a new sequence, we create a fresh instance of our model. We can think about networks that contain recurrent layers by dividing the lifetime of the network instance into discrete time steps. After a certain interval of time, we feed the model the next element of the input. Feedforward connections represent information flow from one neuron to another where the data being transferred is the computed neuronal activation from the current time step. Recurrent connections, however, represent information flow where the data is the stored neuronal activation from the previous time step. Thus, the activations of the neurons in a recurrent network represent the accumulating state of the network instance. The initial activations of neurons in the recurrent layer are parameters of our model, and we determine the optimal values for them just like we determine the optimal values for the weights of each connection during the process of training. It turns out that, given a fixed lifetime (say t time steps) of an RNN instance, we can actually express the instance as a feedforward network (albeit irregularly structured). This clever transformation, illustrated in Figure 12.16, is often referred to as "unrolling" the RNN through time. Let's consider the example RNN in the figure. We'd like to map a sequence of two inputs (each dimension 1) to a single output (also of dimension 1). We perform the transformation by taking the neurons of the single recurrent layer and replicating them t times, once for each time step. We similarly replicate the neurons of the input and output layers. We redraw the feedforward connections within each time replica just as they were in the original network. Then we draw the recurrent connections as feedforward connections from each time replica to the next (since the recurrent connections carry the neuronal activation from the previous time step) (Buduma, 2017).

We can also now train the RNN by computing the gradient based on the unrolled version. This means that all of the backpropagation techniques that we utilized for feedforward networks also apply to training RNNs. We do run into one issue, however. After training each batch example we use, we need to modify the weights based on the error derivatives we calculate. In our unrolled network, we have sets of connections that all correspond to the same connection in the original RNN. The error derivatives calculated for these unrolled connections, however, are not guaranteed to be (and, in practice, probably won't be) equal. We can circumvent this issue by averaging or summing the error derivatives over all the connections that belong to the same set. This allows us to utilize an error derivative that considers all of the dynamics acting on the weight of a connection as we attempt to force the network to construct an accurate output (Buduma, 2017).

12.5 ON THE FUTURE OF NEURAL NETWORKS

Can anyone say what will be the unique role of neural networks and deep learning in all this? To find the answer to these questions, let's take a look at history. During the 1980s there was a lot of excitement and positive attitude towards neural networks, mainly after backpropagation became famous. That enthusiasm decreased, and during the 1990s ML expertise turned to other techniques, mainly support vector machines. In the present, the neural networks are once again flying high, adjusting every form of survey, conquering each and every problem. Can one say that there will come a new technology, far better than neural networks, that would eliminate neural networks and their roles? Or that the benefits that we are getting from the neural networks will come to an end and will not be in use anymore? To answer these points, it would be beneficial to bring into light the concept of ML, rather than neural networks alone. One major aspect is that there is some gap in understanding neural networks. There must be some reason as to why the abstraction of neural network is done so well. The workings of the stochastic gradient descent with its full efficiency has to be examined. The question here arises, how well do the neural networks act? Let's take a short example. If ImageNet was to be multiplied by a factor of 10, how would that effect the performance of neural networks? Would the total efficiency increase or fall down? These are a few basic questions. But at present we know very little in order to answer these questions. Similarly comes the difficulty of predicting the role that neural networks will perform for the betterment of ML. But, according to our understanding so far, we can interpret that ML will never be eradicated. The strength and the ability to analyze numerous stages of hierarchy, creating a number of stages of abstraction, looks to be fundamental. That does not bring us to the conclusion that the process of deep learning in the future would be the same as today. There are a lot of changes being made to the associated units as well as in the architecture of the deep learning algorithms (Nielsen, 2015).

In this chapter the main objective was on using neural nets to do particular tasks, such as distinguishing different pictures. Now, we increase our motivation, and ask how to consider common uses of thinking machines? Could deep learning benefit us in solving the problems related to general-purpose AI? If so, given the increasing popularity of deep learning, could we see common AI in the near future? Answering these questions wisely would lead us to a separate research topic. As an option, we provide one for analysis. Its foundation lies in a concept known as Conway's law: Any system that is designed by an organization would unavoidably produce a design whose architecture is a copy of the organization's communication architecture. Let us take an easy example and consider that an organization is building software for the organization with ML to be incorporated in it, and then they have to take on a person who has expertise in ML algorithms. Conway's law is merely that observation. On knowing Conway's law numerous people reciprocated "Oh… Well! Isn't that banal and so very obvious?" or "Isn't that invalid?" Let's start with the position that Conway's law is wrong and consider all the products that contain meaningless, confusing features or advancements or all the products that have similar critical limitations, e.g., a useless user interface. Now we obtain depth into the history of medical sciences. Long ago, medication was the area of expertise of people like

Galen and Hippocrates, who analyzed the whole body. However, as our intelligence increased, humans were compelled to enhance their knowledge. We found out that what we don't yet see is lots of well-developed subfields, each advancing their own sets of deep ideas, pushing deep learning in several directions. Thus, according to the statistics of social complications, deep learning is, if you'll eliminate the play on words, still a rather shallow field. There is still the possibility of a person learning the deep aspects of the topic (Nielsen, 2015).

12.6 CONCLUSION

So far we know that deep learning is a vast topic and is emerging very rapidly. Numerous ideas can be developed using deep learning, for example imagining things such as the germ theory of disease. For other examples, think of how an antibiotic works or analyzing how the veins, lungs, and most importantly, the heart make the entire cardiovascular system. Such deep analysis shaped the premise for subfields like medicine and other numerous cardiovascular systems. Adding to it the architecture of our information has formed the system of drugs.

REFERENCES

Apostolou, N., King, R. E. (1999 May). "Neuronal state feedback learning of Cohen–Grossberg networks." International Journal of Circuit Theory and Applications 27(3): 331–338.

Ben-David, S. S.-S. (2014). Understanding Machine Learning. Cambridge, MA: Cambridge University Press.

Bishop, C. M. (2006). Pattern Recognition and Machine Learning. Chapter 5: Neural Networks. Springer.

Buduma, N. (2017). Fundamentals of Deep Learning. Sebastopol, CA: O'Reilly.

Kriesel, D. (2005). A Brief Introduction to Neural Networks. Germany.

Nielsen, M. A. (2015). "Neural networks and deep learning." Neural networks and deep learning. Vol. 2018. San Francisco, CA: Determination Press.

Schmid Huber, J. (2015). "Deep learning in neural networks: An overview." Neural Networks. 61: 85–117.

Thomas, D. A. (2017). "An introduction to neural networks for beginners." Technical Report in Adventures in Machine Learning.

13 Machine Learning for Big Data Analytics, Interactive and Reinforcement

Ritwik Raj and Anjana Mishra
C.V. Raman Global University

CONTENTS

13.1 INTRODUCTION

The term "big data" is defined as the large amount of data that is increasing epidemically. It is a way of collecting and organizing large sets of data. Data analytics is the process of examining the sets of data in order to extract the necessary information from the data by building all possible relations among various data. With this the big data seems to appear much bigger than it is. Machine learning (ML) is an application of artificial intelligence (AI) that provides the system the potential to learn on its own from the experiences observed. The volume of data that is termed big data is lot of information for a person to examine, so ML as a service in data analytics has helped in managing this huge amount of data so that it can be processed easily.

13.2 BIG DATA

The big data trend promises to modify our living, working, and thinking by providing access to a mode of optimization, authorizing insight discovery, and improving decision-making. The ability to perceive such huge data depends on how the data has been drawn out using data analytics

13.2.1 CHARACTERISTICS OF BIG DATA

The characteristics of big data are as follows (Figure 13.1):

1. Volume: We know that data is extracted from a large number of places both online and offline, so it's a huge amount of information to keep and maintain precisely. The name "big data" itself tells about volume. The more and the better quality of data, the better is going to be the analysis of it. However, many times storing these huge volumes of data is not easy.
2. Velocity: Velocity is how fast data is transmitted. It deals with calculating how swiftly the data is being transmitted from different sources in the real world, online and offline. The amount of data is very big for large organizations.

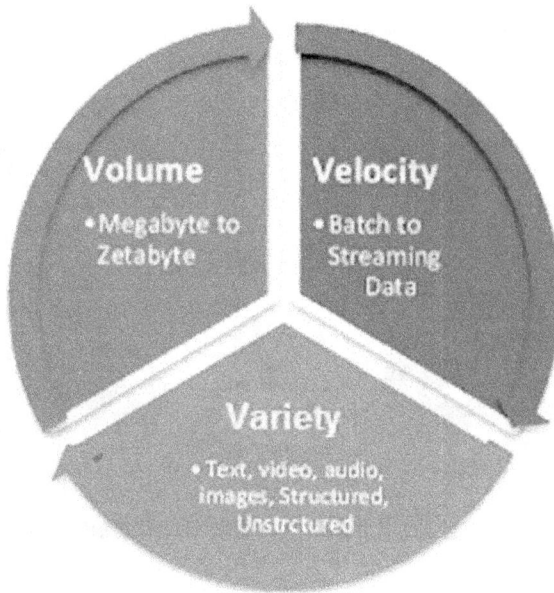

FIGURE 13.1   Characteristics of big data.

3. Variety: Data is visible to us in many forms like text, images, videos, document records, audio, and online sources. Data about the different types available to us tells about the varieties of data (Gunjan Dogra, 2018; Shweta Mittal, Om Prakash Sangwan, 2019).

13.2.2 THE BIG DATA REVOLUTION

Figure 13.2 shows the feeling which has been seen about big data in the past decade. Concealed within the tremendous quantity of amorphous junk data are scraps of information that can allow us to conclude user preferences, trends in the world, and other precious information that helps businesses grow. But why has big data become vitally important very quickly? The presence of data was always there but it was not so easily available to everyone as it is now and it was not present in such a great amount that it would matter. Now, as the Internet has become a very huge part of our lives, we unknowingly leave our impressions everywhere in technological form all over the network. Just like a police detective would excavate clues from a crime scene, different organizations analyze these various technological impressions, making a profile that allows them to compute our habits and know exactly what we want to do at a particular time .Emerging organizations use big data from social media web-browsing, industry predictions, and existing customer records to examine, visualize, and base their business selections. Different organizational sectors use big data to solve their business challenges, transform their processes, and bring about innovation. Retail, medical, banking, government, and security sectors use the big data strategy of trust, availability, and speed to

FIGURE 13.2　Big data revolution.

improve profits. Hence there has been a rush to hire data analysts who use complex languages to analyze millions of pieces of data to make future products or tap into the internal human resources data of multiple companies to assess employee involvement and retain their ability. Big data is among the special ones that have the capability to develop the entire look of tech as it is known to us. Becoming a part of this revolution will help us by shaping our future as we are now aware that big data is the upcoming future (Shweta Iyer, 2019).

13.2.3 WHY IS BIG DATA ANALYTICS IMPORTANT?

The main function of big data analytics is to find solutions to problems such as cost reduction, saving time, and bringing down the risks of decision-making. Organizations are profiting by combining the ML and data analytics:

1. Managing risk and calculating potential causes of risk.
2. Determining causes of business failure and removing them in future.
3. Continuously offering users options in accordance with their purchasing style and preference.
4. Identifying any fraudulent activity by cross-checking data.

13.2.4 CHALLENGES FACED BY BIG DATA

Big data is facing a lot of challenges. Working with big data has become a normal part of business, but this doesn't mean that big data can be handled easily (Figure 13.3). It is quite understandable that one difficulty associated with big data is simply storing and reviewing the data. The amount of information kept in online storage devices

FIGURE 13.3 Challenges of big data.

is doubling at the rate of every two years. In the near future, the total amount will fill a pile of tablets that will reach from the earth to the moon 6.6 times. Enterprises are responsible for about 85 percent of that data. A lot of data doesn't reside in a database because it is unstructured. Analyzing and searching through documents, photos, audio, videos, and other unstructured data can be difficult. To handle the increasing amount of data organizations are developing other options. When storage is concerned, converged and hyper-converged infrastructure and software-defined storage could make it easy for organizations to scaling out their hardware. Costs related to big data storage and the space associated with it can be reduced using technologies like compression, duplication, and tiering. Enterprises are taking the help of various equipment like NoSQL databases, Hadoop, Spark, big data analytics software, business intelligence applications, AI, and ML to help them go deeply through the data. Big data seems to be a very appealing target for hackers, so security is also a big focus of companies who have big data stores. (Cynthia Harvey, 2017)

The goals associated with big data:

1. Decreasing expenses through operational cost efficiencies.
2. Establishing a data-driven culture.
3. Creating new avenues for innovation and disruption.
4. Accelerating the speed with which new capabilities and services are deployed.
5. Launching new product and service offerings.

13.3 MACHINE LEARNING

ML is the field in which computers are capable of learning by themselves from experience without being explicitly programmed. ML is the most wonderful technology of recent generations that has ever come across. The machines are capable of learning by observing and analyzing using ML.

13.3.1 Types of Machine Learning

The main types of ML are:

1. Supervised ML
2. Unsupervised ML

Other types of ML are (Figure 13.4):

- Semi-supervised ML
- Reinforcement learning (mostly considered as supervised learning) (Wikipedia; https://en.wikipedia.org/wiki/Reinforcement_learning)

13.3.1.1 Supervised Learning

In supervised learning we are have a complete chart of the data and we know how the output must look. We have complete knowledge and control over the result and the process is continuously being supervised (Figure 13.5).

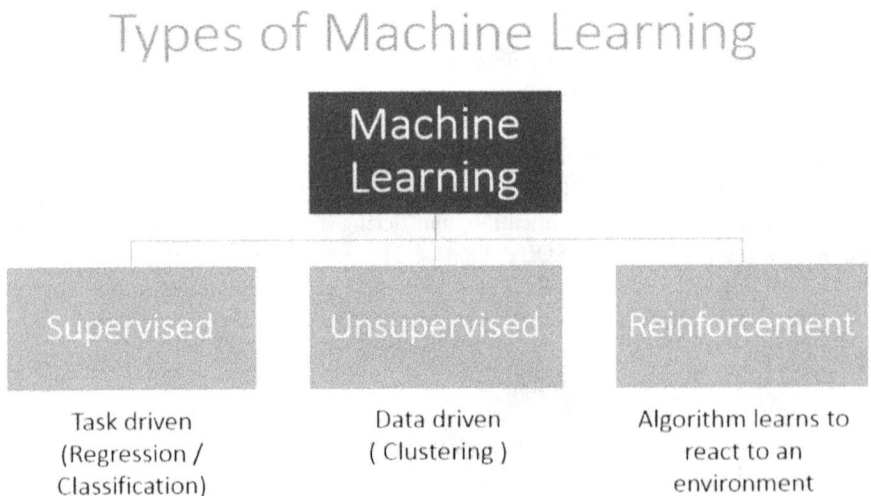

FIGURE 13.4   Classification of ML.

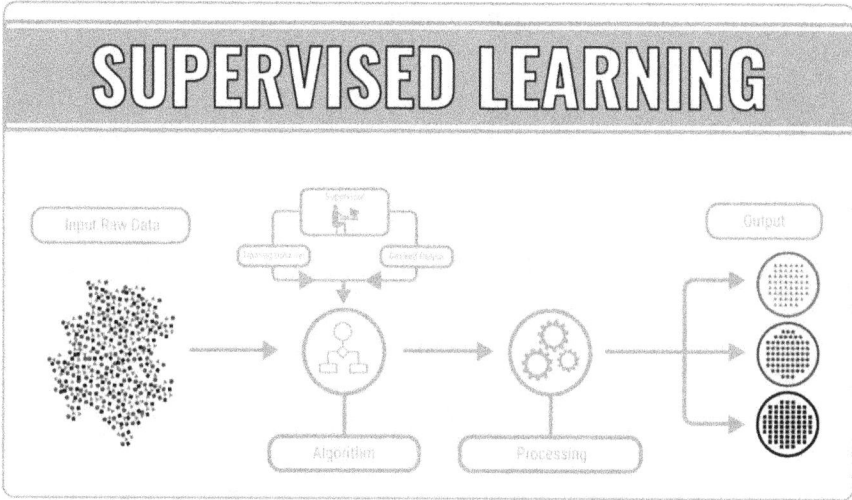

FIGURE 13.5 Supervised learning.

13.3.1.2 Unsupervised Learning

Unsupervised learning is quite similar to supervised, however we have little or no idea about how our output will likely appear, unlike as in supervised learning. We can only draw a rough idea from the structure of the data. We get the structure by combining the data based on relationships among the variables in the data (Figure 13.6).

NOTE: The act of grouping a collection of objects in such a manner that the objects in similar groups (known as clusters) are much more alike (in some sense) to all the objects than those in different groups (clusters) is known as clustering.

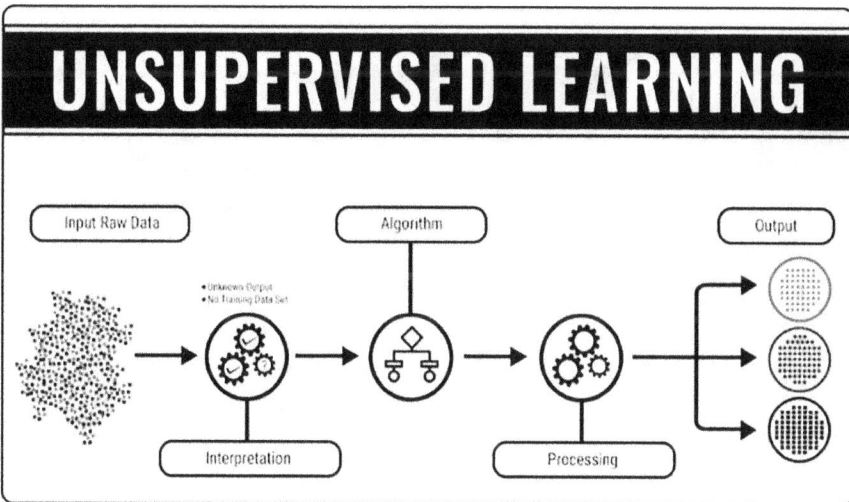

FIGURE 13.6 Unsupervised learning.

Semi-Supervised Learning

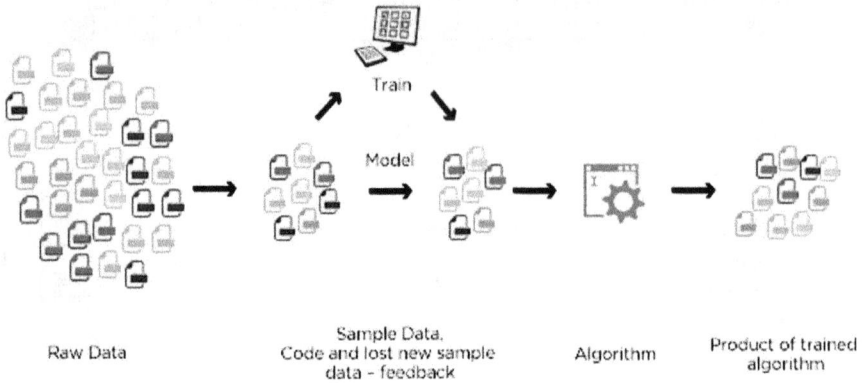

FIGURE 13.7　Semi-supervised learning.

13.3.1.3　Semi-Supervised Learning

ML problems fall into this category when we have very few labeled variables and most of the target variables are unlabeled. We take the help of those few labeled targets to decide classes for the unlabeled targets (Figure 13.7).

13.3.1.4　Reinforcement Learning

Reinforcement learning is taking suitable actions to maximize the benefits one can get in a specified task. It is achieved using specified powerful software and different machines that help to find the best result or path out of any task (J. Qui, Q. Wu, G. Ding, Y. Xu, S. Feng, 2016) (Figure 13.8).

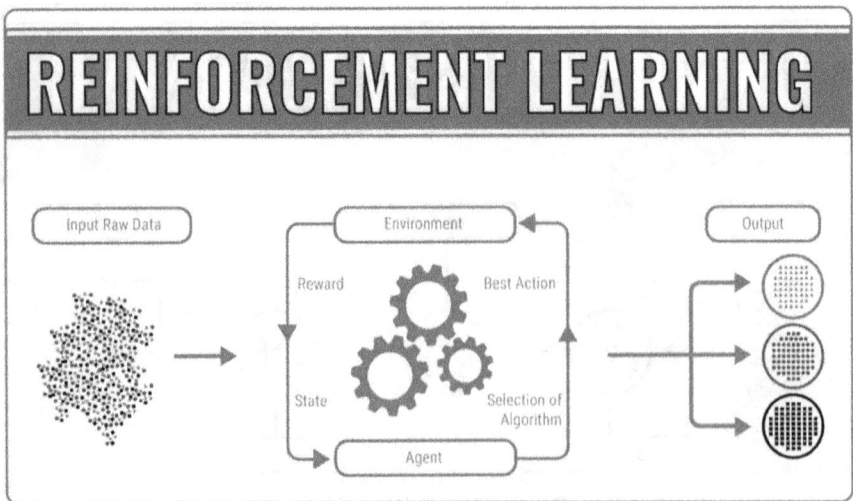

FIGURE 13.8　Reinforcement learning.

13.3.2 MACHINE LEARNING STEPS

13.3.2.1 Collection of Data

- Accuracy of the model is checked by the quantity and quality of data.
- The result of this part is simply a presentation of the information that will be used for training purposes.
- Using data previously collected, through datasets from Kaggle, UCI, etc., now also takes place in this step.

13.3.2.2 Preparing Data

- Dispute data and prepare it for training.
- Randomize data, which erases the effects of the particular order in which we collected and/or otherwise prepared our data.
- Visualize data to help detect relevant relationships between variables or class imbalances (bias alert!), or perform other exploratory analysis.
- Divide data to training and evaluation groups.

13.3.2.3 Selecting a Model

- Choosing the right algorithm is necessary as different algorithms are made for different purposes.

13.3.2.4 Training Model

- The aim of training is to provide an accurate result to the problem.
- Linear regression example: the algorithm would need to learn values for m (or W) and b (x is input, y is output).
- Training step is part of every iteration.

13.3.2.5 Evaluation of the Model

- Uses some bar or combination of bars to "compute" the individual performance of the model.
- Testing the prototype with past data that we have not seen.

13.3.2.6 Tuning of Parameter

- This step refers to hyperparameter tuning, which is an artform as opposed to science. It tunes models for enhanced performance..
- Increase in performance of the tuned model.

13.3.2.7 Making Predictions (Matthew Mayo, 2018)

13.4 APPLICATIONS OF MACHINE LEARNING

The use of AI can be seen anywhere. It is very possible that you are unknowingly using it in your daily routine. A very common application of AI is ML, in which different software operating systems work by perception (just like the human brain). Some examples of ML are used in our daily life without necessarily knowing that they operate via ML. ML has vast applications in real life. There are many day-to-day scenarios

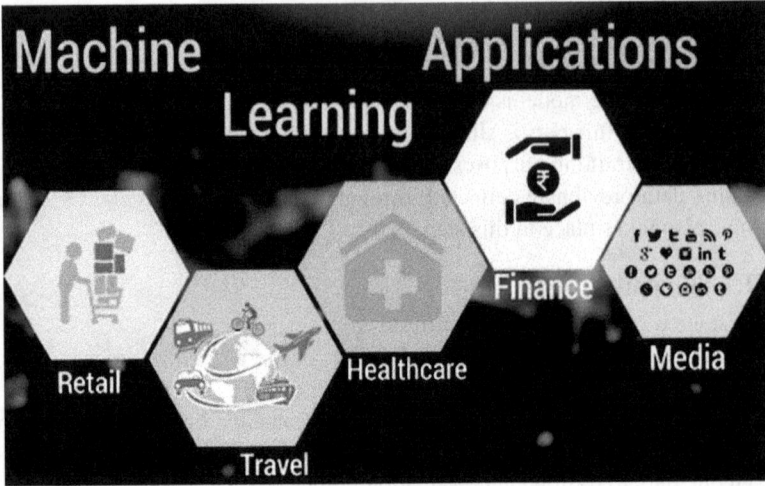

FIGURE 13.9 ML applications.

where ML is applied without the knowledge of the layman user. Determining the relevance of data is the job of ML; the machine is programmed in such a way that it is able to identify the relevance of data by using a variety of sources such as the web address, the number of users frequently browsing particular web pages, how often such web pages appear in similar searches, and the frequency of occurrence of the search string in the entire web page. All this information combines to help the machine learning algorithms (MLAs) understand how relevant a particular web page is for a given search (Intro Books #450, Machine Learning) (Figure 13.9).

13.4.1 VIRTUAL ASSISTANTS

A virtual assistant is like a person who provides various services from a nearby location. Assistants used in smartphones, such as Bixby, Siri, Google Assistant, and Amazon Alexa, are a few of the most popular helpful personal assistants. They assist us by providing information when operated using one's registered voice. All we have to do is switch them on and give commands or questions such as, "What are my appointments today?" "How is the weather in Paris?" or similar questions. To get you the results, your assistant looks through the data, memorizes your searched problems and their solutions, or send an instruction to other apps on your devices to gather information. We could even ask the assistant to perform some work such as, "Set a reminder for tomorrow's meeting" or "Remember to visit Granny." ML is an essential part of these types of virtual assistants because they gather and filter out data based on your previous activities with them. These virtual assistants are connected to a number of different items. A few of them are:

- Smart speakers: e.g., Amazon's Echo Dot and Google's Nest Mini
- Voice assistants: e.g., Samsung's Bixby, Apple's Siri
- Smartphone applications: e.g., Google Assistant

13.4.2 PROGNOSIS

Traffic situations: We usually have GPS access in our devices and we use it very commonly for our convenience. When we use the GPS in our devices we allow companies to track our current speed and location, and all this data is collected at a main server. The information is then processed to make an overview of the latest traffic conditions. These help to control traffic and prevent traffic jams, but the main problem is that there are not many cars or vehicles that have access to a GPS by themselves. ML in these situations is beneficial to find the places where overcrowding is seen based on daily experiences.

Transport networks: Transport networks include taxis, rideshares, and public transit, for example. When we look for a taxi or a cab, now an app calls one and even provides the fare. When these services are shared, the question is how they could lessen the deflection? The solution is ML. The engineering lead at Uber, Jeff Schneider, disclosed that they use ML to explain price rush hours by determining customers demand. Now ML plays a vital role in our lives.

13.4.3 VIDEOS SUPERVISION

Let's think of a person operating numerous surveillance cameras! Such a hectic job. That is why the prospect of upgrading PCs or other devices to perform this job makes sense. The video inspection strategies these days are built with AI that helps to find problems before they occur. They check out negative behavior like standing at a place continuously for a long time, tripping over, or taking nap while sitting etc. The AI detects the activity and sends a report about that particular person, which helps to avoid any kind of issues. When such activities are submitted and found to be true, it helps them to further improve the monitoring. These all work together with the help of ML as a back support internally without coming into consideration of the operators.

13.4.4 SOCIAL MEDIA ACTIVITIES

- *Targeted advertisements:* Making the newsfeed better by properly targeting advertisements, these sources are making use of ML either for purposes of their own or to help users. This is one of the well-known real-world examples of ML that you face on daily basis, without knowing that you see these fabulous things because of ML.
- *People You Should Know:* The concept of ML is to learn from previous experiences. One the best-known things about Facebook, which does a great deal of ML, is that it constantly monitors the activities of people, such as where do we go, who do we meet, what are our field of interests, where do we work, or who do we like to hangout with. After analyzing all these things, Facebook shows us a bunch of people who also have similar tastes as we do.
- *Face Recognition:* Suppose you are uploading a pic of your family and you find that Facebook very quickly identifies the people in the

picture. Facebook locates the views and style in the pic uploaded, sees their uniqueness, and then it matches with each person in our friend list. This process seems to us as very easy but the truth is that all these tasks are very complicated and only made easy by the application of ML.

13.4.5   SPAM E-MAIL AND FILTERING OF MALWARE

A lot of techniques are used by people to identify spam e-mails. Resolving this issue means identifying spam files, so they are continuously upgrading their systems, which are equipped with ML. When junk filtering is completed the process must be repeated, as the junk file creators adopt the latest techniques to produce these mails. Multilayer perceptron and C4.5 decision tree induction are two of the junk-detecting methods that are equipped with ML.

13.4.6   SUPPORTING CUSTOMERS ONLINE

There are numerous online sites these days that offer the choice of communicating with a representative for support. Since we know that every site doesn't has a live person to solve our issues, we find that most of the times you interact with a chatbot. The role of these things is to provide the customer with information taken from different websites. Chatbots have advanced a lot. They have started understanding the problems of the users better and provide truly good results, which has been made possible because of a MLAs approach.

13.4.7   SEARCH ENGINE RESULT REFINING

There are numerous search portals like Google that uses ML to improve search outcomes. Every time a person performs a search, the ML keeps a track of our searching history and how one replies to the results of the search. For example, if someone has opened some of best results and is on the website for a long time then the search engine thinks that the result which it has displayed is right for the searched problem. Similarly, if you keep on going to subsequent pages without opening previous ones then then the search engine thinks that the results didn't match the needs of the person and are inappropriate for the searched problem.

13.4.8   ONLINE FRAUD DETECTION

ML can also solve online fraud by always improving its software through constant evaluation of tasks. It shows its full power when making the Internet a completely protected platform. For example, ML is used by a lot of payment apps like PayPal for privacy and money transaction protection. The organizations use a lot of equipment to enable them to see and provide a comparison between numerous money transfers that are happening and distinguish among the various transactions which are recognized or unrecognized among different people.

13.5 BIG DATA USING MACHINE LEARNING

In today's rapid, growing environment the amount of data gathered by companies is increasing daily. And even though the volume of data that is being collected is not important, the most important task is how the companies are using such a huge amount of data and making profit from it. The streaming of both structured and unstructured data from every corner of the globe at an enormous rate, establishing connections, and retrieving insights is a very complicated task that can swiftly spiral out of control. ML is based on algorithms that can learn from facts without depending on rules-based programming. Big data is the form of data that may be passed into the analytical system so that a ML model could "memorize" (or in other words, improve the accuracy of its prognostics).We often talk about ML and big data in the same flow but they are not the same things. ML is required to extract the best information out of big data.

13.5.1 ENTER MACHINE LEARNING

Modern businesses know that big data is powerful, but they're starting to realize that it's not nearly as useful as when it's paired with intelligent automation. With the increase in demand for ML, soon the price of AI will decrease, helping the whole world to adopt AI. AI machines have to be trained so that they can easily walk and communicate with people from different cultures and backgrounds. With massive computational power, ML systems help companies manage, analyze, and use their data far more successfully than ever before. Here's how multinational companies across industries are using big data technology to grow long-term business value.

13.5.2 ML AND BIG DATA—REAL-WORLD APPLICATIONS

Machine learning: The branch of AI that gave us self-driving cars has been useful in analyzing bigger, very complexed data to see invisible patterns, explore markets, and analyze the preferences of the customer for faster, more accurate results. In big data, ML is providing an interconnection of machines with huge databases to make them learn a lot of new things all by themselves. Big companies are analyzing big data with the help of MLAs to predict future trends in the market.

Healthcare: ML capabilities are impacting healthcare in profound ways, by improving diagnostics and personalizing treatment plans. Predictive analysis enables doctors and clinicians to focus on providing better service and patient care, creating a proactive framework for addressing patient needs before they are sick. Big data and ML help us to see the signs of disease, which helps us to identify problems rapidly. It will also help in developing new medications. The management of medical data, such as previous health records, reports from labs, etc., also becomes easier with the help of this. In return the data provides a clear view of the health status of any person. Wearable technologies and sensors are also available to assess the patient's health in real time, detecting trends or red flags that could potentially foresee a dangerous health event such as cardiac arrest. Advancements in cognitive automation can support a diagnosis by quickly analyzing large volumes of medical and healthcare data, identifying patterns, and connecting the dots to enhance treatment and care.(P. Y. Wu, C. W. Cheng, C. D. Kaddi, J. Venugopalan, R. Hoffman, M. D. Wang, 2017).

Retail: In retail, relationship-building is critical for success. The ML-powered technologies collect, examine, and work on that data to simplify the experiences of real-time shopping. The algorithms uncover similarities and differences in customer data to accelerate and make the segmentation simpler for better targeting. ML may have helped improve the accuracy of the workforce, but this is not entirely true. Since it is very true that the machines lack emotions and don't feel any sentiments, they always require human involvement that can check the market conditions in different ways. Machines are only able to work as well as the algorithms are designed by us. However, based on learned preferences, deeper analysis can push undecided shoppers toward conversion. For example, ML abilities can help customers who are shopping online by providing proper recommendations of necessary products with reasonable prices, vouchers, and other real-time offers. With customer experience top of mind, Walmart is working to develop its own proprietary ML and AI technologies. In March of 2017, the retail chain opened Store №8 in Silicon Valley, a dedicated space and incubator for developing technologies that will enable stores to remain competitive in the next five to ten years.

Financial services: In the financial sector, predictive analytics help prevent fraud by analyzing large historical data sets and building forecasts based on previous data. ML models learn behavior patterns and then — with little human interaction — anticipate events for more informed decision-making. Now, by using big data and machine learning, we can reach the world trade market without distress. Banks and financial institutions use ML to gather real-time insights that help drive investment strategies and other time-sensitive business opportunities.

Automotive and other industries: In the face of stiff competition, the automotive industries are starting to leverage ML strength and big data analytics to make operations, marketing, and customer experience better before, during, and after the purchase of products. Applying statistical models to historical data helps automakers identify the impact of past marketing efforts to define future strategies for improved return on investment. This method of prediction helps producers and dealers check and explore sensitive data in regard to part failures, minimizing the cost of maintenance for customers. The network of dealers can be optimized by tracking for accurate, real-time parts inventory and improved customer support. As ML technologies hit new levels of maturity, smart businesses are shifting their approaches to big data. Across industries, companies are reshaping their infrastructures to maximize intelligent automation, integrating their data with smart technologies to improve not only productivity, but also their ability to better cater to their customers.

13.5.3 IMPLEMENTING MACHINE LEARNING IN BIG DATA

MLAs provide efficient, self-sufficient tools for data storage, analysis, and integration. If one's organization is not very big and there's not a lot of information coming it, ML wouldn't be required because all the tasks could be done using a basic set of tools or manually. On the other hand, when one has an organization that deals with huge data, MLAs help to utilize time efficiently and effectively. Together with cloud computing benefits, ML allows rapid and precise analysis and summation of many datasets, whether they concern user behavior, sales, or DNA sequencing. Machines learn better the more data they have at their disposal. Big data analytics thus gives machines the volume and variety of data they need to make increasingly better and more efficient decisions in the performance of tasks. It makes sense — a veteran basketball player with a bigger and more varied "dataset" of experience will usually play better than a rookie.

The MLA could be implemented with many forms of big data operations:

- Data labeling/segmentation
- Data analytics
- Descriptive
- Diagnostic
- Predictive
- Prescriptive
- Planning
- Scenario simulation

Combining these elements allows users to see the big picture, created from big data with patterns, insights, and all other items of interest sorted out, categorized, and packaged into a digestible form. It's important to understand that ML applied to big data results in an infinite loop. The creation of certain algorithms for specific tasks is being watched and upgraded over time as the data comes and goes through the machine (Volodymyr Bilyk, 2019).

13.5.4 EMPOWERING BIG DATA AND MACHINE LEARNING

As the world grows daily at an enormous rate, the size of data, collectively called big data, also grows exponentially. At the same time, another revolution taking the world to another level is going on in the field of technological enhancement, that is ML and AI. We can say in simple language that ML is a collection of equipment that empowers linked machinery and computers to learn, evolve, and improve by restating and frequently checking stored data and by continuously investigating human development. As far as the reality is concerned, technical giants, organizations, and data scientists all over the world are working on big data to make a big difference in the ML and AI world. In a recent survey conducted for big data executives by New Vantage Partners, about 88.5% of top executives believe that AI is very soon going to be seen as the biggest influence that might create a threat for their companies (Stevenn Hansen, 2019) (Figure 13.10).

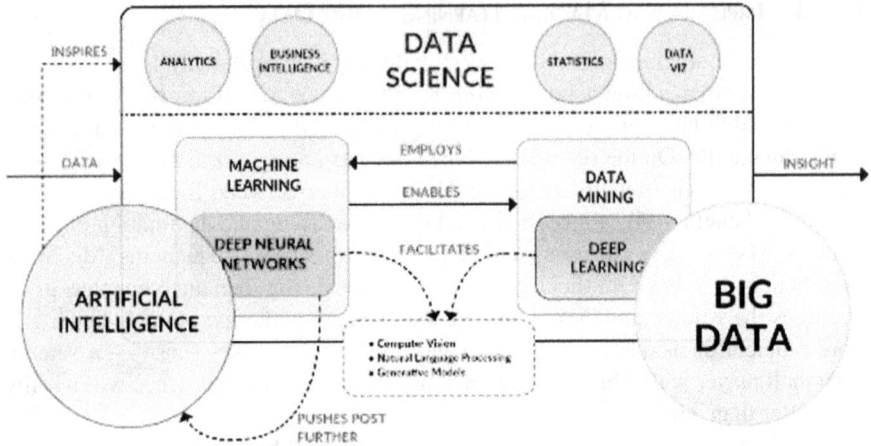

FIGURE 13.10   Empowerment of big data with ML.

(Image Courtesy: Whatsthebigdata)

13.6 CONCLUSION AND THE FUTURE OF BIG DATA ANALYTICS

As data sets continue to grow and produce more real-time streaming data, businesses are turning to alternative storage option like cloud storage, etc. The continuous exchange of data is changing the way we live in society. The huge impact of this bulk amount of data cannot be passed over. The amount of progress and developing changes is having both a direct and indirect impact on us; these new discoveries are going to lead to huge amounts of data in the near future and managing this data will also become necessary. Since the powerful and accurate evaluation of data has provided us with the power and the ability to make decision in businesses, now more developed systems are needed to fulfill the requirements of future. The MLA provide huge support to handle big data. Numerous businesses are becoming knowledgeable about the difficulties of working with a wide range of information. It is quite clear that big data and ML are going to be very important in the near future and it must be handled very carefully so that it can be easily used by anyone.

REFERENCES

Volodymyr Bilyk, Oct 25, 2019, The App Solutions, https://theappsolutions.com/blog/development/machine-learning-and-big-data/.

Gunjan Dogra, May 15, 2018. Characteristics of Big Data, Indian National Interest, Sep28, 2019, https://nationalinterest.in/big-data-analytics-using-machine-learning algorithms-c33ef8488638.

Stevenn Hansen, Oct 14, 2019, Big Data Empowering, Oct 26, 2019, https://hackernoon.com/how-big-data-is-empowering-ai-and-machine-learning-4e93a1004c8f.

Cynthia Harvey, June 5, 2017, Big Data Challenges, Datamation, Oct 25, 2019, https://www.datamation.com/big-data/big-data-challenges.html.

Shweta Iyer, Oct 1, 2019, Big Data Revolution, knowledgehut, Oct 26, 2019, https://www. knowledgehut.com/blog/big-data/the-big-data-revolution.

Matthew Mayo, May, 2018. ML Steps, KDnuggets, Oct 10, 2019 https://www.kdnuggets. com/2018/05/general-approaches-machine-learning-process.html.

Shweta Mittal, Om Prakash Sangwan, "Big Data Analytics using Machine Learning Techniques," 9th International Conference on Cloud Computing, Data Science & Engineering (Confluence), Noida, India, 2019, pp. 203–207, doi: 10.1109/ CONFLUENCE.2019.8776614.

J. Qui, Q. Wu, G. Ding, Y. Xu, S. Feng, "A survey of machine learning for big data processing", EURASIP Journal on Advances in Signal Processing, vol. 2016, no. 67, pp. 1–16, 2016.

P. Y. Wu, C. W. Cheng, C. D. Kaddi, J. Venugopalan, R. Hoffman, M. D. Wang, "ML and big data-real world applications 'omic and electronic health record big data analytics for precision medicine'", IEEE Transactions on Biomedical Engineering, vol. 64, no. 2, pp. 263–273, 2017.

14 Fish Farm Monitoring System Using IoT and Machine Learning

Farjana Yeasmin Trisha and
Mohammad Farhan Ferdous
Japan-Bangladesh Robotics & Advanced
Technology Research Center (JBRATRC)

Mahmudul Hasan
Jahangirnagar University

CONTENTS

14.1 INTRODUCTION

Fish cultivation is an especially important business around the world. Individuals sometimes cultivate fish on their lake or pond, and now and then they develop a fish farm on their housetop or on their yard. Yet, at times they endure a major misfortune in this business, when the fish get sick. They need to learn about fish-farming conditions, what conditions are great or awful for fish, and how to boost the fish's development in aquaculture, lake, or pond. What's more, sufficient learning can increase their profitability. They need to know which components are required to augment the water condition. The main principle of our project is to monitor the water environment so that we can choose what step to follow next. Here we use different sensors. We measure temperature, turbidity, pH, water level, and a gas sensor. Using the sensors with a robotized technique we send data to a Web platform so we can see it. We use Ethernet Shield to send data to our Internet of Things (IoT) platform. The IoT server stores the data in a cloud server for future analysis. By this method we can see the current condition of that water at a particular moment.

We can see and compare water environmental conditions at any time and whatever point. Fish are naturally cold blooded (Mulchandani, Marshettiwar and Varma 2019). So, if the temperature is increasing past a specific point, fish will die. Notwithstanding whether the fish can withstand a general water temperature increase, any sudden, marked change in water temperature will significantly affect fish physiology. So, we ought to worry about water temperature. Low water temperature decreases the fish's metabolic rate and increases the amount of dissolved oxygen in the water (Fatani, Kanawi and Alshami 2018).

The pH level is very important for fish. In the event that the pH level drops below 4.5, fish will die. In general, in a fish farm, we should maintain acidity and alkalinity of the water between 6–8 for better outcomes and fish growth.

Fish can be divided into three types based on where they live in the water: upper-level, mid-level, and low-level. Diverse sorts of fish only survive in their water level so water level is critical for fish surviving. Different scale of fish can survive in different positions. Below their maximum level fish can't survive. On the other hand, if the water level is too low the fish that prefer the low levels can't survive. So water level preference varies for different types of fish. Low-level fish can't get by in a shallow situation.

At the deepest level of the lake, very little grass and other phytoplankton grow. On the off chance that they grow too much, they die. After they die, they make carbon monoxide and ammonia gas expands. If they increase too much then they rot. When small leaves and other small plants rot they produces different types of gases like carbon monoxide, ammonia, and many other harmful gases. These gases are harmful for fish. They can't survive in that environment. So if that type of gas increases in the water environment that means we need to clean the rotten grass below water.

When turbidity levels increase, dissolved oxygen will decline and daylight can't go through the water. Fish can't find proper level of oxygen. Also, the water temperature decreases in the lower level of the water. So fish farmers should stay aware of the turbidity. If we check the water quality of a fish farm, we should concentrate on these types of essential elements.

14.2 RELATED WORK

Mulchandani, Marshettiwar and Varma (2019) study how fish are cold-blooded animals, regulating their body temperature directly by the water environment. Changes in water temperature influence the amount of dissolved oxygen in the water and fish oxygen utilization. In spite of the fact that the fish can withstand a wide water temperature range, any sudden, outrageous changes in water temperature will considerably affect fish physiology. A chilling injury will cause cold shock to the fish, leading to death or paralysis with a loss of balance. The reason might be the respiratory center, or osmotic regulation is influenced at excess temperatures.

Myint, Gopal and Aung (2017) research the IoT condition, where a sensor interface device is basic for information collection in wireless sensor network (WSN) system for water quality monitoring (WQM) as the water contamination is a critical issue universally. The system contains a collection of WQM sensors like temperature, pH, turbidity, etc. In this project they use XBee to receiving sensor value. They use FPGA protocol to collect the sensor value using XBee. This paper introduces the structure of a WSN-based, reconfigurable smart sensor interface device for a WQM system in an IoT situation. The execution of the proposed reconfigurable WSN system is checked through computer simulation and research center.

Fatani, Kanawi and Alshami (2018) study the potential of hydrogen level (pH) plays a significant role in fish farming, industries, and agriculture. In this paper, a tool is built up by using Arduino that is connected to both a designed mobile app and a physical device using the Internet. By using the IoT concept, this paper creates a tool that allows real-time control of the pH level of a specific product or solution. This research collects and controls pH value measurements.

Vijayakumar and Ramya (2015) study a real-time monitoring system designed to ensure the safe supply of quality drinking water. They develop a low-cost system for real-time monitoring of the water quality using the IoT concept. A different type of sensors is used in this system that can be viewed on the Internet using cloud computing. The Raspberry Pi B+ model is used here as a core controller.

Prathibha, Hongal and Jyothi (2017) propose that IoT will assume a crucial role in digital agriculture. Monitoring environmental factors are the main factor to enhance the yield of productive crops. A camera is interfaced with CC3200 to capture pictures and send the images through multimedia messaging service (MMS) to the farmer's mobile device using Wi-Fi. The element of this paper includes checking temperature and humidity in a farming field through sensors utilizing CC3200 single chip.

Moparthi, Mukesh and Sagar (2018) implement the system of the drinking water reservoir and municipal water tanks by using an Arduino board for finding the pH value and a GSM module for message technique. Here they use a LED display for continuous observation of water parameters. By collecting the data of the pH value of water they extend the system by sending sensor data to the cloud platform for global monitoring of water quality.

Kumar and Jasuja (2017) proposed a model for air quality monitoring to detect air pollution of our environment. He found different types of gas values like carbon dioxide, carbon monoxide, and other important gases by using Raspberry Pi.

14.3 PROPOSED SYSTEM MODEL

This chapter presents a novel approach to investigate these issues, which represent an obstacle toward the deployment of fish-farming systems. The fish cultivating framework can be observed utilizing Web availability. The real obstructions to organizing fish cultivation dependent on IoT are the robotization of the framework where the sensors will be interfaced. Following that, working with the framework and using it daily has introduced some interesting research avenues the intrigued scientists to pursue. The fundamental advantage of the framework is that it provides a robotized interface for the sensors that can be observed remotely from an alternate area through the Web. This is a unique way to collect data remotely. Figure 14.1 shows the hardware architecture diagram of the system.

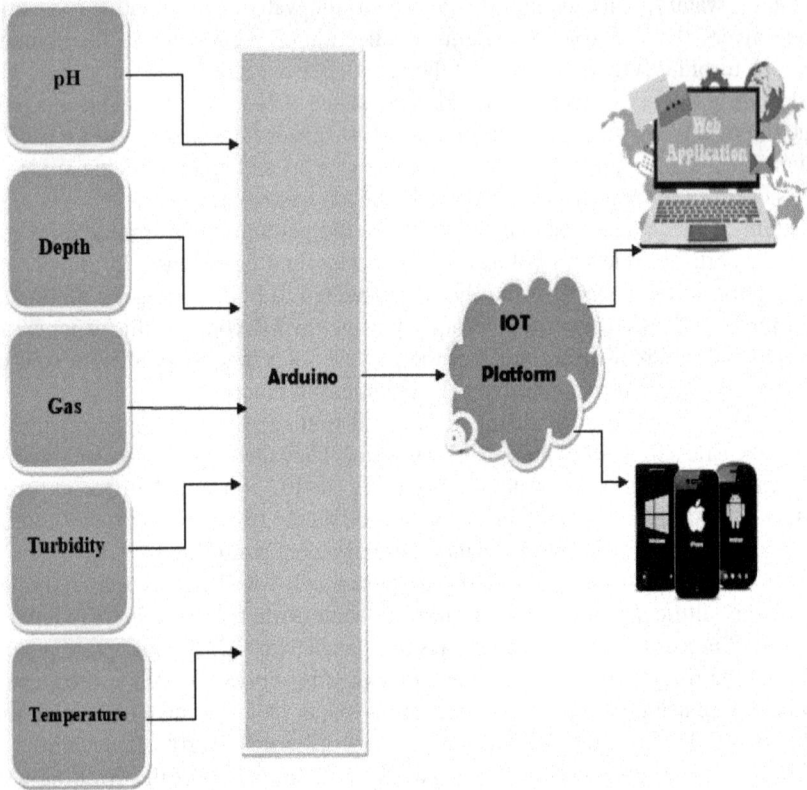

FIGURE 14.1   Proposed model.

Flow Chart of the Following System:
Figure 14.2 is the system flowchart of the system. Start following the process step by step and complete all processes. After summing all the procedure we will find our expected device.

14.4   HARDWARE SECTION

A. **Arduino Uno:** For collecting all sensor data together we use an Arduino Uno board. Arduino Uno is a microcontroller board in perspective of the ATmega328P (datasheet). It has 14 modernized information/yield pins (of which 6 can be used as PWM yields), 6 straightforward data sources, a 16 MHz quartz crystal, a USB affiliation, a power jack, an ICSP header, and a reset get. It contains everything anticipated to help the microcontroller; just interface it to a PC with a USB connection or power it with an AC-to-DC connector or battery to start (Figure 14.3).

START

Combined All Sensor

Connect All Sensors to
the Internet

Store Data on Server

Show on a Platform

Analysis Data

Result

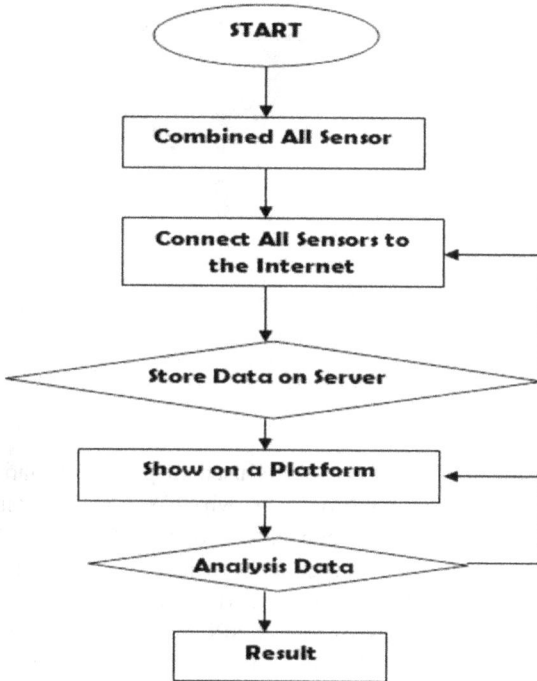

FIGURE 14.2 System flowchart.

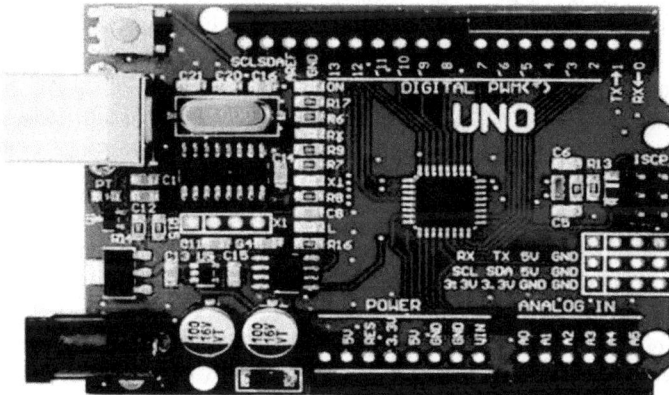

FIGURE 14.3 Arduino Uno.

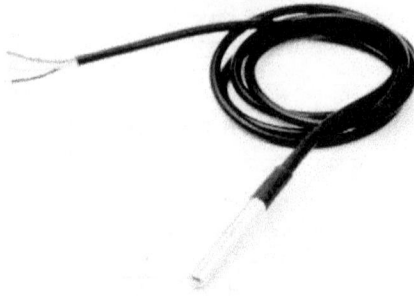

FIGURE 14.4   DS18B20 temperature sensor.

B. **Ethernet Shield:** Ethernet Shield is used to pass information from the Arduino board to the Web through the wired Web association. Just plug it into the Arduino board and connect with Internet by RJ45 cable. All sensors are combined and associated on the Arduino board then it is associated with Ethernet Shield. Arduino and Ethernet Shield have same pin configuration. It connects with Arduino on an SPI port. When data passes through Ethernet Shield its LED light will blink continuously.

C. **DS18B20 Temperature sensor:** A waterproof temperature sensor used with the help of DS18B20 one wire temperature sensor. We can measure the temperature from –55°C to 125°C with an accuracy of ±5. DS18B20 is called one wire programmable temperature sensor (Figure 14.4).

D. **pH sensor:** A pH meter is a coherent instrument that appraises the hydrogen-particle development in water-based arrangements, demonstrating its alkalinity or acidity introduced as pH. A pH meter has two basic segments, one with a pointer that moves against a scale or a computerized meter. The second segment is a pH sensor which is an advanced meter. We make a circuit board and it's joined to an Arduino, which has some code for working with a pH meter. pH is an extent of acidity or alkalinity of the solution, the pH scale ranges from 0 to 14 (Figure 14.5).

E. **Turbidity sensor:** To detect water quality we need to measure water turbidity. For measuring water turbidity we use turbidity sensor SKU: SEN 0189. This sensor provides both analog and digital signal outputs. The analog condition is used for its threshold. We interface the water turbidity sensor with Arduino Uno to measure water turbidity. Figure 14.6 shows the turbidity sensor of the system. The turbidity sensor can detect different types of water turbidity and give accurate values.

F. **Water level sensor module (ultrasonic sensor):** To measure the distance between the water surface and ground level of water we use a sonar ultrasonic sensor for perfect measurement. An ultrasonic transmitter is utilized by the water level detecting module to radiate to an ultrasonic recipient.

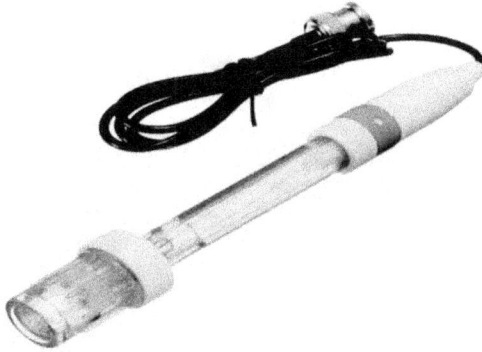

FIGURE 14.5   pH sensor.

The sound wave reflection time changes depending on the water level. The intensity of the transmitted acoustic wave is influenced by distance attenuation, which is the result of the energy scatter from the developing distribution area. Estimating distance based on the time contrast as a reference for precise determination of the echo, the frequency of measurement ought to be increased to get a higher resolution so that the distance information is more accurate.

G. **Ammonia and carbon monoxide gas sensor:** To distinguish ammonia gas we use MQ-7.It can detect carbon monoxide gas anywhere from 20 to 2000ppm. The MQ-135 sensor is suitable for detectingNH_3, NO_x, alcohol, benzene, smoke, CO_2, etc. Here we need to detect ammonia gas, so we use the MQ-135 sensor (Figure 14.7).

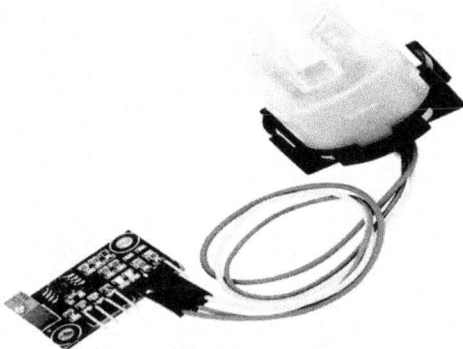

FIGURE 14.6   Turbidity sensor.

FIGURE 14.7   MQ-135 gas sensor.

14.5 EXPERIMENTAL SETUP

Figure 14.8 demonstrate our trial setup and implanted framework respectively. Here, we connect all the sensors together. We assemble the circuit and get information by running Arduino code. Then we connect Ethernet Shield to send the data to a platform. We use API key to connect the IoT platform and our Arduino board. After connecting, all data will be stored in the server. We can see our real-time data on the server dashboard.

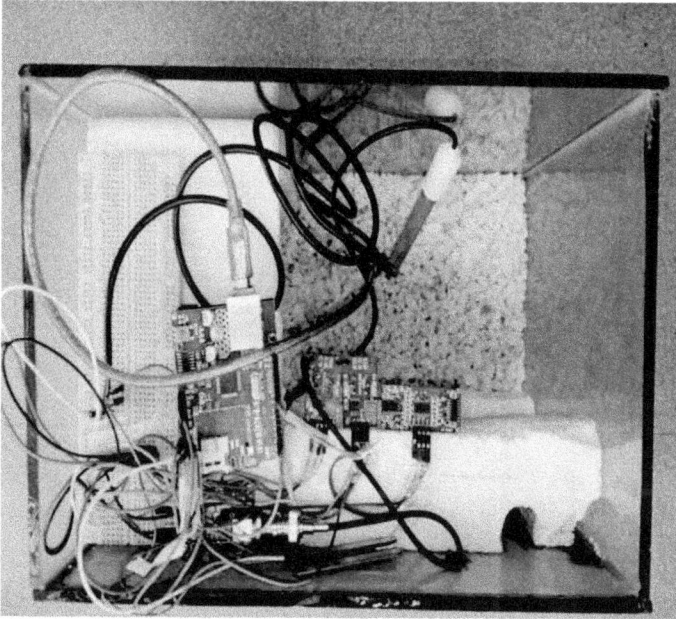

FIGURE 14.8   Experimental setup.

First we connect Ethernet Shield with the Arduino board. Arduino and Ethernet Shield have the same pin configuration. So connect those two things top to bottom. After connecting Arduino and Ethernet Shield, we need to set up all sensors with Ethernet Shield one by one. Connect all sensors in analog and digital output pins. After connecting all sensors we use Arduino IDE to generate Arduino code to find all sensor's data. Then we need to send it to an IoT platform to store information for future data processing and analysis. We implement the system in a rural area of Bangladesh and find outstanding results. It can be implemented anywhere in the world with a very low cost. There is no need to monitor this system regularly in person.

14.6   RESULT

According to the proposed plan, we have executed the equipment as shown in Figure 14.8 and utilizing embedded C coding in Arduino IDE. Figure 14.8 illustrates the corresponding results obtained by processing the data received from different sensors such as temperature, pH level, turbidity, water level, and ammonia and other gases. Arduino and Ethernet Shield both connect together where all these signals are processed.

Here Figure 14.9 is our expected result that we find after simulating Figure 14.7 we find all sensor values together in Arduino com port. Figure 14.9 is Arduino com port dashboard. When we run Arduino code the values will show continuously in this dashboard. Here the first value is the most recent value. The value of the level of water is measured directly. Other parameters like pH and turbidity must still be calibrated.

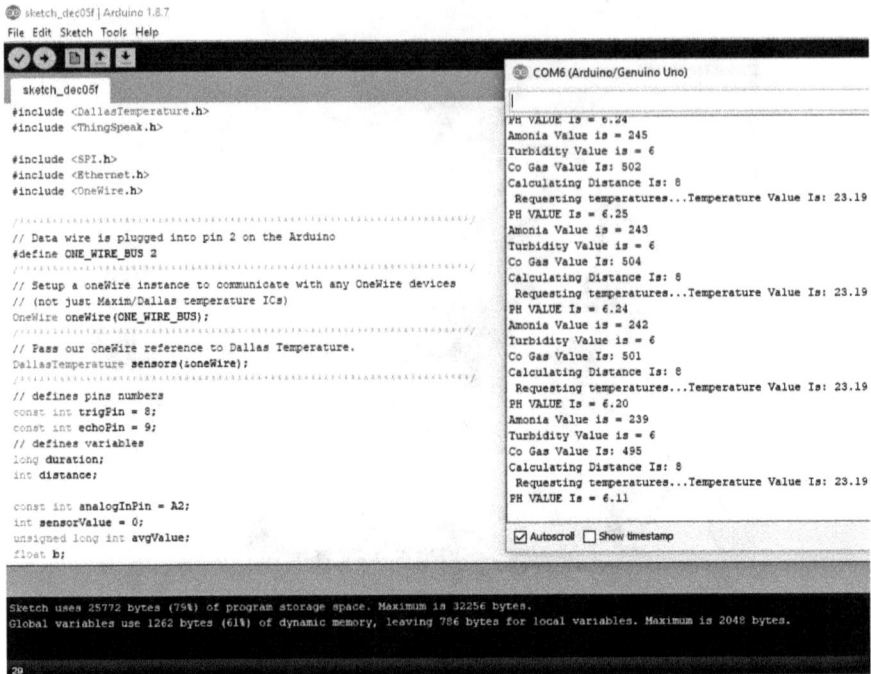

FIGURE 14.9 Arduino code.

pH sensor changes over the pH level into the corresponding voltage value, subsequently measuring the best possible voltage we determine the real pH value. If the pH value is 0 to 6, the water is acidic, from 8 to 14 it is basic, and at a pH value of 7 it is neutral. Hence, as indicated by these ranges, we decide whether the water is acidic, basic, or neutral.

Turbidity can be evaluated from the voltage level we have gained and measured in unit NTU. We got the voltage, measured in NTU, and the pH value.

After receiving all sensor information we store the data in a database. From there, we accumulate real information on fisheries which are valuable for fishery operators (Figure 14.10).

At that point, we apply naive Bayes algorithm in our collecting dataset and train the dataset by machine learning algorithm. After gathering all the sensor information we store the data in a database, that is, we collect a real-time dataset on various fisheries in Bangladesh. Then we prepare and process our data and apply different algorithms. We find the best result by applying the naive Bayes algorithm. In Table 14.1 (below) we can see the precision, recall, and F-measure rate.

We can also calculate accuracy by following rules. We apply it to our collecting dataset. After training the algorithm we test our collecting dataset which we find by using the sensor, and then test it. The machine thus predicts whether our environment is good or harmful for fish. Applying MLA from our collected real-time dataset we find a decision that is very accurate.

```
    COM6 (Arduino/Genuino Uno)

Agrofisheries Using IOT
Amonia Value is = 272
Turbidity Value is = 7
Co Gas Value Is: 500
Calculating Distance Is: 8
 Requesting temperatures...Temperature Value Is: 23.12
PH VALUE Is = 6.21
Amonia Value is = 269
Turbidity Value is = 7
Co Gas Value Is: 502
Calculating Distance Is: 8
 Requesting temperatures...Temperature Value Is: 23.12
PH VALUE Is = 6.29
Amonia Value is = 268
Turbidity Value is = 7
Co Gas Value Is: 503
Calculating Distance Is: 8
 Requesting temperatures...Temperature Value Is: 23.12
PH VALUE Is = 6.29
Amonia Value is = 265
Turbidity Value is = 7
Co Gas Value Is: 503
Calculating Distance Is: 8
 Requesting temperatures...Temperature Value Is: 23.12
PH VALUE Is = 6.28

☑ Autoscroll  ☐ Show timestamp                              Newline
```

FIGURE 14.10   Arduino code.

We apply the dataset to our real-time project and get an outstanding result that reduces the wastage costs of fish farmers. Figure 14.11 shows the result when we use data analysis software to analyze our dataset, applying the naive Bayes algorithm to our dataset because the naive Bayes algorithm predicts more accurately than other algorithms.

Figure 14.12 shows our platform where we can see our dashboard and can visualize our experiment value.

We can visualize real-time data anytime and anyplace. We can also analyze our water environment so that we can decide what we should do if any alteration happens. Here in Table 14.1 we can see the detailed information after applying machine learning algorithms to our collecting dataset.

TABLE 14.1

Finding Dataset after Machine Learning Algorithms

Summary	Results
Correctly Classified Instances	98.1308%
Incorrectly Classified Instances	1.8692%
Kappa Statistics	0.9125
Mean Absolute Error	0.0226
Root Mean Squared Error	0.1372
Relative Absolute Error	9.7088%
Root Relative Squared Error	40.6653%

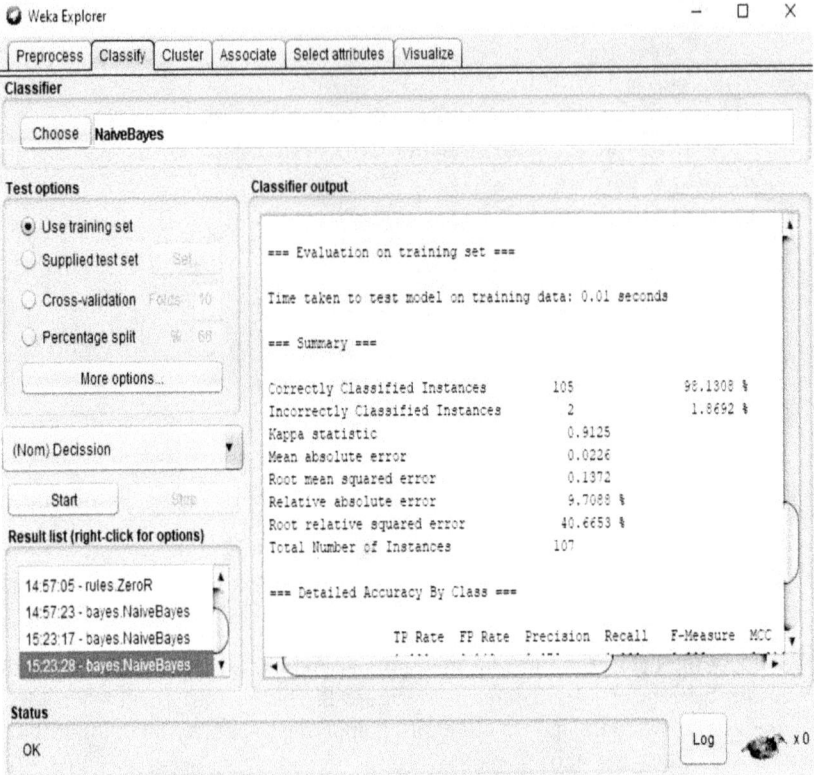

FIGURE 14.11   Data analysis.

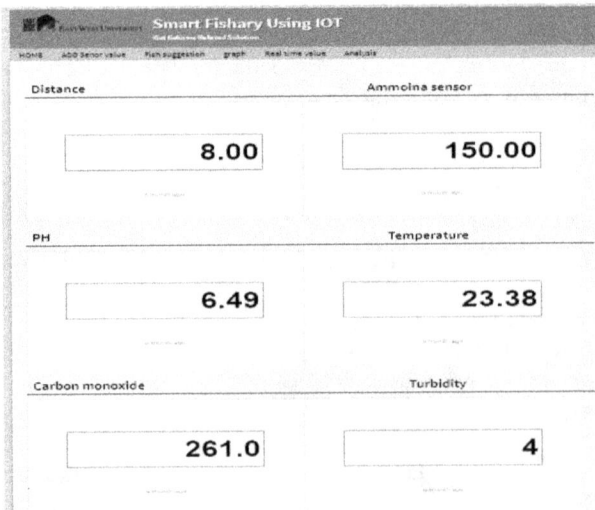

FIGURE 14.12   Our platform.

TABLE 14.2
Machine Learning Model Prediction

TP Rate	FP Rate	Precision	Recall	F-Measure	PRC Area	Class
1.000	0.143	0.979	1.000	0.989	0.920	Good
0.857	0.000	1.000	0.857	0.923	0.920	Harmful
0.981	0.124	0.1241	0.982	0.981	0.920	

Here we calculate that TP, FP rate, precision, and recall in our dataset and find its accuracy by using an equation:

$$Accuracy = \frac{TP + TN}{TP + TN + FP + FN}$$

Where TP, TN, FP, and FN denote true positives, true negatives, false positives, and false negatives, respectively. According to recall and precision shown in Table 14.2, the accuracy of the rule induction model was 98.13%.

14.7   CONCLUSION

This is very efficient and cost-effective approach for fish farmers where they can measure the water environment using this architecture. We can measure real-time data by utilizing those sensors and send that information by an Ethernet Shield via Internet connection and send those real-time data on IoT platform. From here we can observe it anytime and anyplace and can take the necessary steps. On the off chance that the agriculturist uses this innovation in their cultivation they can expand their efficiency. They can monitor the water environment so they can take immediate steps when a vital component changes. Currently they cannot know the conditions right then and there. To find a good environment for the fish farm we apply MLA called naive Bayes so that we can predict whether our present conditions are good or not. This setup can reduce productive cost and get more benefit. It can also increase our efficiency.

14.8   FUTURE WORK

To start, we don't have a major aquarium or lake to measure natural equipment perfectly. We also require a dissolved oxygen meter for estimating oxygen. However, this is too expensive for us and for this project. We also require a phytoplankton detecting disk to estimate phytoplankton to give us an approximate idea of accessible fish food. We plan to collect a bigger dataset and implement it on Python code to increase accuracy and visualization. We emphatically believe that when we assemble this equipment we can give the best results for cultivating fish. We will implement these features very soon to make our project a standard and business cycle project. The fish farmer will especially appreciate that the possibility of failure will drop by 1–5%. We also maximize our efficiency.

REFERENCES

1. M. Mulchandani, D. Marshettiwar, R. Varma. "Autofish monitoring system." International Journal of Computer Sciences and Engineering, vol. 7, no. 12, 133–138, May 2019.
2. Asma Fatani, Afnan Kanawi, Hedaih Alshami. "Dual pH level monitoring and control using IoT application." 15th Learning and Technology Conference. Jeddah, Saudi Arabia: IEEE, 2018, 167–170.
3. Cho Zin Myint, Lenin Gopal, Yan Lin Aung. "WSN-based reconfigurable water quality monitoring system in IoT environment." 14th International Conference on Electrical Engineering/Electronics, Computer, Telecommunications. Phuket, Thailand: IEEE, 2017, 741–744.
4. Vijayakumar, R. Ramya. "The real time monitoring of water quality in IoT environment." International Conference on Innovations in Information, Embedded and Communication systems. Coimbatore, India: IEEE, 2015, 1–5.
5. S.R.I. Prathibha, Anupama Hongal, M.P. Jyothi. "IoT based monitoring system in smart agriculture." International Conference on Recent Advances in Electronics and Communication Technology. Bangalore, Karnataka, India: IEEE, 2017, 81–84.
6. Nageswara Rao Moparthi, Ch. Mukesh, P. Vidya Sagar. "Water quality monitoring system using IoT." International Conference on Advances in Electrical, Electronics, Information, Communication and Bio-Informatics. Chennai, India: IEEE, 2018, 1–5.
7. Somansh Kumar, Ashish Jasuja. "Air quality monitoring system based on IoT using raspberry Pi." International Conference on Computing, Communication and Automation. Greater Noida, India: IEEE, 2017, 1341–1346.
8. Y.M. Poonam, Y. Mulge. "Remote temperature monitoring using LM35 sensor and intimate android user via C2DM service." International Journal of Computer Science and Mobile Computing, vol. 2, no 6, 32–36, 2013.
9. P. Raju, R.V.R.S. Aravind, S. Kumar. "Pollution monitoring system using wireless sensor network in Visakhapatnam," International Journal of Engineering Trends and Technology (IJETT), vol. 4, 591–595, April 2013.
10. R.V.P. Yerra, M.S. Baig, R.K. Mishra, R. Pachamuthu, U.B. Desai, S.N. Merchant. "Real time wireless air pollution monitoring system," ICTACT Journal on Communication Technology, vol. 2, no. 2, 370–375, June 2011.
11. Maneesha V. Ramesh, Nibi K.V. Anupama Kurup, Amrita Vidyalayam. "Water quality monitoring and waste management using IoT." Global Humanitarian Technology Conference. San Jose, CA, USA: IEEE, 2017, 1–7.
12. Zexin Lin, Weixing Wang, Huili Yin, Sheng Jiang. "Design of monitoring system for rural drinking water source based on WSN." International Conference on Computer Network, Electronic and Automation. Xi'an, China: IEEE, 2017. 289–293.
13. S. Usha Kiruthika, S. Kanaga Suba Raja. "IoT based automation of fish farming." Journal of Advance Research in Dynamical & Control Systems, vol. 9, no. 1, 50–56, 2017.
14. M. Garcia, S. Sendra, G. Lloret, J. Lloret. "Monitoring and control sensor system for fish feeding in marine fish farms." IET Communications (IET), vol. 5, no. 12, 1682–1690, September 2011.
15. Wang Xiaoyi, Dai Jun, Liu Zaiwen, Zhao Xiaoping, Dong Suoqi, Zhao Zhiyao, Zhang Miao. "The lake water bloom intelligent prediction method and water quality remote monitoring system." 2010 Sixth International Conference on Natural Computation (ICNC 2010). Yantai, China: IEEE, 2010, 3443–3446.
16. Jui-Ho Chen, Wen-Tsai Sung, Guo-Yan Lin. "Automated monitoring system for the fish farm aquaculture environment." International Conference on Systems, Man, and Cybernetics. Hong Kong: IEEE, 2015, 1161–1166.

17. Jieying Xiao, Zijing Guo. "Detection of chlorophyll-a in urban water body by remote sensing." Second IITA International Conference on Geoscience and Remote Sensing. Qingdao, China: IEEE, 2010, 302–305.

18. Muhammad Bilal, Abdullah Gani, Mohsen Marjani, Nadia Malik. "A study on detection and monitoring of water quality and flow." 12th International Conference on Mathematics, Actuarial Science, Computer Science and Statistics (MACS). Karachi, Pakistan: IEEE, 2018, 1–6.

19. S. Anjana, M.N. Sahana, S. Ankith, K. Natarajan, K.R. Shobha, A. Paventhan. "An IoT based 6LoWPAN enabled experiment for water management." International Conference on Advanced Networks and Telecommuncations Systems. Kolkata, India: IEEE, 2015, 1–6.

20. Li Shijin, Zhu Haichen, Chen Deqing, Wang Lingli. "Water quality monitoring based on multiple remote sensing imageries." Fourth International Workshop on Earth Observation and Remote Sensing Applications. Guangzhou, China: IEEE, 2016, 1–5.

21. Zihan Pang, Kebin Jia. "Designing and accomplishing a multiple water quality monitoring system based on SVM." 2013 Ninth International Conference on Intelligent Information Hiding and Multimedia Signal Processing, Beijing, China: IEEE, 2013, 121–124.

Index